Global Warming Book!

Malcolm B. Riel

Copyright and Legal Disclaimer:

Copyright © 2024 by Malcolm B. Riel

All rights reserved. No part of this book may be reproduced, stored, or transmitted in any form or by any means, electronic, mechanical, photocopying, recording, scanning, or otherwise, without the prior written permission of the publisher, except as permitted under the Copyright Act of 1976.

This book is a work of non-fiction, intended to provide information and insights into the complex topics of climatology, environmental science, geology, and weather. While the author, Malcolm B. Riel, has made every effort to ensure the accuracy of the information presented herein, readers are advised to independently verify any information or data before making decisions or taking actions based on the contents of this book.

The views and opinions expressed in this book are those of the author and do not necessarily reflect the views of the publisher or any affiliated individuals or organizations.

The information provided in this book is for educational and informational purposes only and is not intended as professional advice. Readers are encouraged to consult with qualified professionals in relevant fields for specific guidance or assistance pertaining to their particular circumstances.

Neither the author nor the publisher shall be held liable for any damages, losses, or injuries arising from the use of or reliance on the information presented in this book, whether directly or indirectly.

Every effort has been made to secure permissions for all copyrighted material used in this book. Any omissions or oversights are unintentional, and the author and publisher will rectify them in subsequent editions if notified.

For permissions inquiries or any other questions regarding this book, please contact the publisher at the address provided in the front matter.

Thank you for reading "Global Warming Book." Your engagement with this important subject matter contributes to a better understanding of the challenges and opportunities we face in addressing the global climate crisis.

Malcolm B. Riel

"Global Warming Book"

Chapter: Introduction 9

Chapter: The Author's Perspective and Purpose 11

Chapter: Overview of the Book's Structure 13

Chapter: The Fundamentals of Climate Science 15

Chapter: Climatology: Exploring the Earth's Climate Systems 17

Chapter: Historical Perspectives on Climate Change 19

Chapter: The Role of Greenhouse Gases 21

Chapter: Natural Climate Variability vs. Anthropogenic Influence 23

Chapter: Environmental Science: Understanding Ecosystem Dynamics 25

Chapter: Impacts of Global Warming on Biodiversity 27

Chapter: Ocean Acidification and its Consequences 29

Chapter: Deforestation and Carbon Sequestration 32

Chapter: Geology: Unraveling the Earth's Geological History 34

Chapter: Geological Evidence of Past Climate Change 36

Chapter: Impact of Human Activities on Geomorphology 38

Chapter: Geological Solutions to Climate Challenges 41

Chapter: The Science Behind Global Warming 44

Chapter: Atmospheric Processes and Weather Patterns 47

- Chapter: The Greenhouse Effect and its Mechanisms 50
- Chapter: Extreme Weather Events and Climate Change 53
- Chapter: Predictive Modeling and Climate Projections 56
- Chapter: Oceanic Dynamics: The Heartbeat of the Planet 59
- Chapter: Thermal Expansion and Sea Level Rise 62
- Chapter: Ocean Currents and Climate Regulation 65
- Chapter: Feedback Loops and Tipping Points 68
- Chapter: The Human Factor: Understanding Anthropogenic Contributions 71
- Chapter: Fossil Fuel Combustion and Carbon Emissions 74
- Chapter: Land Use Changes and Urbanization 77
- Chapter: Technological Solutions and Mitigation Strategies 80
- Chapter: Impacts and Adaptation 83
- Chapter: Ecological Impacts: Disruptions to Ecosystem Services 86
- Chapter: Loss of Habitats and Species Extinction 89
- Chapter: Food Security and Agricultural Challenges 92
- Chapter: Human Health Risks and Disease Dynamics 95
- Chapter: Socioeconomic Consequences: Addressing Inequality and Vulnerability 98
- Chapter: Displacement and Climate Refugees 101

Chapter: Economic Costs of Climate Change 104

Chapter: Social Justice and Climate Equity 107

Chapter: Adaptation Strategies: Building Resilience in a Changing World 110

Chapter: Sustainable Development and Green Infrastructure 113

Chapter: Community-Based Adaptation Initiatives 117

Chapter: Policy Interventions and International Cooperation 121

Chapter: Looking Forward: Pathways to a Sustainable Future 125

Chapter: The Power of Collective Action: Mobilizing for Change 128

Chapter: Grassroots Movements and Activism 131

Chapter: Political Will and Policy Reform 135

Chapter: Corporate Responsibility and Sustainable Business Practices 139

Chapter: Innovation and Technology: Shaping a Low-Carbon Future 143

Chapter: Renewable Energy Revolution 147

Chapter: Carbon Capture and Storage Technologies 150

Chapter: Advances in Climate Engineering 154

Chapter: Cultivating Hope: Inspiring Hopeful Narratives and Positive Change 157

Chapter: Education and Public Awareness Campaigns 160

Chapter: The Role of Art and Culture in Climate Communication 163

Chapter: Conclusion - Recapitulation of Key Points 166

Chapter: Call to Action: Individual Responsibility and Collective Engagement 168

Chapter: The Imperative of Hope: Embracing a Vision of Sustainability 172

Appendix: Glossary of Key Terms 175

Chapter: Introduction

Setting the Stage: The Urgency of Global Warming

As we stand at the threshold of the 21st century, we find ourselves confronted with an existential threat unlike any other in human history: global warming. The Earth's climate is changing at an unprecedented rate, driven primarily by human activities such as the burning of fossil fuels, deforestation, and industrialization. The consequences of these actions are far-reaching and profound, impacting ecosystems, economies, and communities around the world.

In this chapter, we set the stage for our exploration of global warming by examining the urgent need for action in the face of this looming crisis. We begin by delving into the scientific evidence that unequivocally demonstrates the reality of climate change, from rising temperatures to melting ice caps and increasingly frequent extreme weather events. We will also explore the disproportionate impacts of global warming on vulnerable communities, underscoring the importance of addressing climate change from a perspective of social justice and equity.

But perhaps most importantly, we will consider the implications of inaction. The longer we delay meaningful action to mitigate and adapt to climate change, the more severe and irreversible its impacts will become. From coastal flooding and food shortages to mass migration and conflict, the stakes could not be higher.

Yet, even in the face of such daunting challenges, there is reason for hope. The global community has the knowledge, technology, and resources to address climate change and build a more sustainable future for all. By coming together with determination, creativity, and solidarity, we can rise to meet the challenge of global warming and create a world in which both humanity and the planet thrive.

In the pages that follow, we will explore the science, impacts, and solutions to global warming in depth, laying the groundwork for a deeper understanding of this complex and urgent issue. It is my hope that by shining a light on the urgency of global warming, we can inspire action and collective engagement in the pursuit of a sustainable and resilient future.

Let us embark on this journey together, with courage, determination, and a shared commitment to safeguarding our planet for generations to come.

Chapter: The Author's Perspective and Purpose

In this chapter, I, Malcolm B. Riel, offer insights into my perspective and purpose in writing "Global Warming Book." As a seasoned researcher and advocate for environmental sustainability, my journey into the realm of climate science has been driven by a deep-seated passion for understanding and addressing the challenges posed by global warming.

My perspective on global warming is rooted in a profound sense of responsibility — to our planet, to future generations, and to the countless species with whom we share this precious Earth. Over the years, I have witnessed firsthand the devastating impacts of climate change on ecosystems, communities, and livelihoods around the world. From disappearing glaciers to unprecedented heatwaves and catastrophic storms, the signs of our changing climate are all too clear.

But amidst these challenges, I remain steadfast in my belief that we have the power to make a difference. My purpose in writing "Global Warming Book" is twofold: to educate and to inspire. Through meticulous research and a commitment to scientific accuracy, I aim to provide readers with a comprehensive understanding of the science behind global warming. From the fundamental principles of climatology to the intricacies of weather patterns and ocean dynamics, I endeavor to demystify the complexities of climate change and empower readers with knowledge.

Yet, knowledge alone is not enough. It is my hope that by sharing the urgency of the climate crisis and exploring potential solutions, "Global Warming Book" will serve as a catalyst for action. Whether you are a concerned citizen, a policymaker, or a student of environmental science, I invite you to join me on this journey of discovery and advocacy. Together, we can work towards a more sustainable and resilient future for all.

As we embark on this journey, I encourage you to approach the subject with an open mind and a willingness to engage with the complexities of global warming. By coming together with determination, creativity, and solidarity, we can confront the challenges of climate change and chart a course towards a brighter tomorrow.

Thank you for joining me on this important mission. Together, let us strive to safeguard our planet for generations to come.

Malcolm B. Riel

Chapter: Overview of the Book's Structure

In this chapter, I provide readers with an overview of the structure and organization of "Global Warming Book." As we embark on our journey through the complexities of climate change, it is important to understand how the book is structured and how each section contributes to our overall understanding of the issue.

"Global Warming Book" is divided into several key sections, each of which explores a different aspect of the climate crisis in depth. These sections are designed to build upon one another, providing readers with a comprehensive understanding of the science, impacts, and solutions to global warming.

The first section of the book, titled "The Fundamentals of Climate Science," lays the groundwork for our exploration of global warming. Here, we delve into the principles of climatology, environmental science, geology, and weather, examining the factors that contribute to climate change and the evidence that supports our understanding of this complex phenomenon.

Next, in "The Science Behind Global Warming," we take a closer look at the mechanisms driving climate change and the ways in which it manifests in our world. From atmospheric processes and weather patterns to oceanic dynamics and the human factors driving anthropogenic climate change, this section provides readers with a deeper understanding of the science behind global warming.

In the third section, "Impacts and Adaptation," we explore the far-reaching consequences of climate change and the strategies that can help us mitigate and adapt to its effects. From ecological impacts and socioeconomic consequences to adaptation strategies and resilience-building initiatives, this section highlights the urgency of addressing global warming and the importance of taking action.

Finally, in "Looking Forward: Pathways to a Sustainable Future," we explore the potential solutions to the climate crisis and the opportunities for positive change. From collective action and innovation to hopeful narratives and positive examples, this section offers readers a vision of a more sustainable and resilient future.

Throughout "Global Warming Book," readers will find a blend of scientific analysis, real-world examples, and practical guidance. Each chapter is designed to inform, inspire, and empower readers to take action in the fight against climate change.

As we journey through the pages of "Global Warming Book," I encourage readers to approach the subject with an open mind and a willingness to engage with the complexities of the climate crisis. Together, we can confront the challenges of global warming and build a brighter future for generations to come.

Chapter: The Fundamentals of Climate Science

In this chapter, we embark on a journey into the heart of climatology, exploring the fundamental principles that underpin our understanding of the Earth's climate system. From historical perspectives to contemporary research, we delve into the mechanisms that drive climate change and the evidence that supports our understanding of this complex phenomenon.

We begin by examining the historical context of climate change, tracing the evolution of our understanding from early observations to modern scientific research. By studying ancient climate records preserved in ice cores, tree rings, and sediment layers, scientists have pieced together a detailed picture of past climate variations, providing valuable insights into the natural cycles and variability of the Earth's climate.

Next, we explore the role of greenhouse gases in shaping the Earth's climate. From carbon dioxide to methane and nitrous oxide, these gases act like a blanket, trapping heat in the atmosphere and warming the planet. While some greenhouse gases occur naturally, human activities such as the burning of fossil fuels and deforestation have significantly increased their concentrations, leading to rapid and unprecedented warming.

We also investigate the concept of climate variability, examining the natural factors that influence climate patterns on both short and long timescales. From solar cycles and volcanic eruptions to oceanic oscillations such as El Niño and La Niña, these natural factors play a crucial role in shaping regional climate variations and extreme weather events.

But perhaps most importantly, we confront the evidence of human-induced climate change. Through meticulous research and sophisticated climate models, scientists have documented the unmistakable fingerprints of human activities on the Earth's climate system. From the rapid rise in global temperatures to the acidification of the oceans and the melting of polar ice caps, the impacts of human-induced climate change are becoming increasingly clear and undeniable.

As we delve deeper into the fundamentals of climate science, it becomes evident that the Earth's climate is a complex and interconnected system, shaped by a multitude of factors. By understanding the mechanisms driving climate change and the evidence supporting our understanding, we can better appreciate the urgency of addressing this global challenge.

In the chapters that follow, we will explore the science behind global warming in greater detail, examining the atmospheric processes, oceanic dynamics, and human factors driving anthropogenic climate change. By building a solid foundation of knowledge, we can equip ourselves with the tools and understanding needed to confront the challenges of global warming and build a more sustainable future for generations to come.

Chapter: Climatology: Exploring the Earth's Climate Systems

In this chapter, we embark on a journey into the fascinating field of climatology, where we explore the intricate systems that govern the Earth's climate. Climatology is the study of long-term weather patterns, encompassing the interactions between the atmosphere, oceans, land surfaces, and living organisms. By understanding these complex systems, we gain valuable insights into the processes that shape our planet's climate and the factors that drive climate change.

We begin by examining the Earth's energy balance, the delicate equilibrium between incoming solar radiation and outgoing heat energy. The atmosphere plays a crucial role in regulating this balance, trapping heat through the greenhouse effect, and maintaining the Earth's temperature within a narrow range suitable for life. However, human activities such as the burning of fossil fuels and deforestation have intensified the greenhouse effect, leading to a rapid rise in global temperatures.

Next, we delve into the mechanisms that drive climate variability on both regional and global scales. From atmospheric circulation patterns such as the jet stream and trade winds to oceanic currents like the Gulf Stream and the El Niño Southern Oscillation (ENSO), these systems play a key role in shaping weather patterns and climate conditions around the world. By studying these phenomena, climatologists can predict future climate trends and better understand the impacts of climate change.

We also explore the concept of climate feedback loops, where changes in one part of the climate system can amplify or mitigate the effects of climate change. For example, melting polar ice caps reduce the Earth's albedo, or reflectivity, leading to further warming as more sunlight is absorbed by the darker ocean waters. Similarly, the release of methane from thawing permafrost can exacerbate global warming, creating a vicious cycle of climate feedback.

But climatology is not just about understanding the past and present—it's also about predicting the future. By using sophisticated climate models and observational data, climatologists can forecast future climate trends and assess the potential impacts of climate change on ecosystems, economies, and societies. These projections serve as valuable tools for policymakers, allowing them to develop strategies to mitigate and adapt to the effects of global warming.

As we journey through the field of climatology, it becomes clear that the Earth's climate is a complex and dynamic system, shaped by a multitude of factors. By unraveling these complexities and understanding the mechanisms that drive climate change, we can better appreciate the urgency of addressing this global challenge. In the chapters that follow, we will explore the impacts of climate change on biodiversity, ecosystems, and human societies, and discuss strategies for building a more sustainable future for all.

Chapter: Historical Perspectives on Climate Change

In this chapter, we delve into the rich tapestry of Earth's climate history, tracing the evolution of our planet's climate over millions of years. By examining ancient climate records and geological evidence, we gain valuable insights into the natural cycles and variations that have shaped the Earth's climate throughout history.

Our journey begins deep in the past, as we explore the geological archives that offer glimpses into ancient climates. From ice cores drilled from polar ice caps to sediment layers preserved in ocean floors and lake beds, these records provide valuable clues about past climate conditions, including temperature fluctuations, atmospheric composition, and the distribution of ice sheets and glaciers.

One of the most striking findings from these records is the existence of natural climate cycles, which operate on timescales ranging from decades to millions of years. For example, the Earth has experienced numerous ice ages and interglacial periods over the past few million years, driven by variations in the Earth's orbit and axial tilt. These cycles, known as Milankovitch cycles, play a key role in shaping long-term climate patterns and have profound implications for the stability of the Earth's climate system.

We also explore the impact of geological processes such as volcanic eruptions and tectonic activity on past climate variations. Major volcanic eruptions can inject large quantities of ash and sulfur dioxide into the atmosphere, leading to temporary cooling known as volcanic winter. Similarly, tectonic movements can alter the distribution of land and sea, influencing ocean currents and atmospheric circulation patterns, and thus shaping regional climate conditions.

But perhaps the most remarkable aspect of Earth's climate history is the resilience of life in the face of extreme environmental changes. Throughout geological time, organisms have adapted to shifting climate conditions, evolving new traits and behaviors to survive and thrive in changing environments. By studying the fossil record and ancient ecosystems, paleoclimatologists can reconstruct past climates and gain insights into the resilience of life in a dynamic and ever-changing world.

As we reflect on the historical perspectives of climate change, it becomes clear that the Earth's climate is a complex and dynamic system, shaped by a multitude of factors. By understanding the natural cycles and variations that have characterized our planet's climate history, we gain valuable insights into the processes that drive climate change today. In the chapters that follow, we will explore the impacts of human activities on the Earth's climate system and discuss strategies for mitigating and adapting to the effects of global warming.

Chapter: The Role of Greenhouse Gases

In this chapter, we delve into one of the fundamental drivers of climate change: greenhouse gases. These gases play a crucial role in regulating the Earth's temperature and maintaining the conditions necessary for life as we know it. However, human activities have significantly altered the composition of the atmosphere, leading to an intensification of the greenhouse effect and unprecedented warming of the planet.

We begin by exploring the basic principles of the greenhouse effect. When sunlight reaches the Earth's surface, some of it is absorbed and converted into heat energy. This heat is then radiated back towards space in the form of infrared radiation. Greenhouse gases in the atmosphere, such as carbon dioxide (CO_2), methane (CH_4), and nitrous oxide (N_2O), trap some of this outgoing radiation, preventing it from escaping into space and warming the Earth's surface. Without the greenhouse effect, the Earth would be too cold to support life as we know it.

Next, we examine the sources and sinks of greenhouse gases in the Earth's atmosphere. While some greenhouse gases occur naturally, others are produced by human activities such as the burning of fossil fuels, deforestation, and industrial processes. These activities release large quantities of greenhouse gases into the atmosphere, leading to a rapid increase in atmospheric concentrations. Meanwhile, natural processes such as photosynthesis and oceanic uptake act as sinks, removing greenhouse gases from the atmosphere and helping to regulate their levels.

We also explore the concept of radiative forcing, which quantifies the impact of greenhouse gases on the Earth's energy balance. Positive radiative forcing occurs when greenhouse gases trap more heat in the atmosphere, leading to warming, while negative radiative forcing occurs when they allow more heat to escape, leading to cooling. By measuring radiative forcing, scientists can assess the relative contributions of different greenhouse gases to global warming and develop strategies to mitigate their effects.

But perhaps the most alarming aspect of the role of greenhouse gases in climate change is the rapid rate at which their concentrations are increasing. Over the past century, atmospheric concentrations of CO_2 have risen to levels not seen in millions of years, primarily due to the burning of fossil fuels for energy. This unprecedented increase in CO_2 and other greenhouse gases has led to a rapid warming of the planet and unprecedented changes in the Earth's climate.

As we reflect on the role of greenhouse gases in climate change, it becomes clear that human activities are driving profound and far-reaching changes to the Earth's climate system. In the chapters that follow, we will explore the impacts of global warming on ecosystems, societies, and economies, and discuss strategies for mitigating and adapting to the effects of climate change.

Chapter: Natural Climate Variability vs. Anthropogenic Influence

In this chapter, we delve into the complex interplay between natural climate variability and anthropogenic influence, seeking to untangle the factors driving changes in the Earth's climate system. While natural processes have long played a role in shaping the Earth's climate, human activities have increasingly become a dominant force, driving unprecedented changes to the global climate.

We begin by examining the natural drivers of climate variability, which operate on a range of timescales, from months to millions of years. These include factors such as variations in solar radiation, volcanic eruptions, and natural fluctuations in ocean currents and atmospheric circulation patterns. For example, the El Niño Southern Oscillation (ENSO) is a natural climate phenomenon characterized by periodic warming and cooling of sea surface temperatures in the tropical Pacific Ocean, which can have far-reaching impacts on weather patterns around the world.

Next, we explore the concept of climate proxies, which are indirect indicators of past climate conditions. These include tree rings, ice cores, sediment layers, and fossilized pollen, which can provide valuable insights into past climate variability and the natural cycles and trends that have characterized the Earth's climate over geological time scales. By studying these proxies, scientists can reconstruct past climates and gain a better understanding of the natural drivers of climate change.

However, in recent centuries, human activities have increasingly become a dominant force driving changes to the Earth's climate. The burning of fossil fuels for energy, deforestation, industrial processes, and agricultural practices have led to a rapid increase in atmospheric concentrations of greenhouse gases, such as carbon dioxide (CO_2), methane (CH_4), and nitrous oxide (N_2O). These gases trap heat in the atmosphere, leading to a rapid warming of the planet and unprecedented changes in the Earth's climate.

To distinguish between natural climate variability and anthropogenic influence, scientists use a variety of methods, including climate models, observational data, and attribution studies. These studies allow researchers to quantify the relative contributions of natural and human factors to observed changes in the Earth's climate and assess the likelihood of future climate scenarios.

But perhaps the most compelling evidence of human influence on the Earth's climate comes from the observed changes in temperature, precipitation patterns, sea level rise, and other climate indicators. These changes are occurring at an unprecedented rate and are consistent with the expected impacts of human-induced climate change, providing strong evidence that human activities are driving changes to the Earth's climate system.

As we reflect on the complex interplay between natural climate variability and anthropogenic influence, it becomes clear that human activities are driving profound and far-reaching changes to the Earth's climate system. In the chapters that follow, we will explore the impacts of climate change on ecosystems, societies, and economies, and discuss strategies for mitigating and adapting to the effects of global warming.

Chapter: Environmental Science: Understanding Ecosystem Dynamics

In this chapter, we explore the intricate dynamics of Earth's ecosystems and the ways in which they are influenced by climate change. Environmental science provides us with a lens through which we can understand the complex interactions between living organisms and their environment, and how global warming is altering these interactions.

We begin by examining the concept of an ecosystem, a community of living organisms (biotic) interacting with their physical environment (abiotic) within a defined area. Ecosystems range from small-scale habitats like ponds and forests to vast biomes like tropical rainforests and polar regions. Each ecosystem is characterized by unique species compositions, energy flows, and nutrient cycles, which are intricately interconnected and sensitive to changes in environmental conditions.

Next, we delve into the impacts of climate change on ecosystem dynamics. Rising temperatures, shifting precipitation patterns, and more frequent extreme weather events are disrupting ecosystems around the world, altering species distributions, migration patterns, and phenology (the timing of biological events such as flowering and migration). For example, warming temperatures are causing glaciers to melt, permafrost to thaw, and sea levels to rise, leading to habitat loss and fragmentation for many species.

We also explore the concept of ecosystem services, the benefits that ecosystems provide to humans and other organisms. These services include provisioning services (such as food, water, and raw materials), regulating services (such as climate regulation, flood control, and water purification), cultural services (such as recreation, spiritual enrichment, and aesthetic enjoyment), and supporting services (such as nutrient cycling, soil formation, and pollination). Climate change is threatening the stability and resilience of ecosystem services, with potentially profound consequences for human well-being and biodiversity.

But despite the challenges posed by climate change, ecosystems also possess remarkable resilience and adaptive capacity. By studying ecosystem dynamics and the processes that govern them, scientists can identify strategies for enhancing ecosystem resilience and promoting biodiversity conservation in the face of global warming. These include measures such as habitat restoration, sustainable land management practices, and the creation of protected areas and wildlife corridors.

As we reflect on the intricate dynamics of Earth's ecosystems, it becomes clear that they are both vulnerable to the impacts of climate change and vital for mitigating its effects. In the chapters that follow, we will explore the impacts of climate change on biodiversity, ecosystems, and human societies, and discuss strategies for building resilience and promoting sustainability in a changing world.

Chapter: Impacts of Global Warming on Biodiversity

In this chapter, we explore the profound impacts of global warming on Earth's biodiversity, highlighting the threats posed to species and ecosystems around the world. Biodiversity, the variety of life on Earth, is essential for the functioning of ecosystems and provides numerous benefits to human society, including food, medicine, and cultural enrichment. However, climate change is rapidly altering ecosystems, leading to shifts in species distributions, loss of habitats, and increased extinction risks.

We begin by examining the concept of biodiversity and the factors that contribute to its richness and complexity. Biodiversity encompasses three main components: genetic diversity (the variety of genes within species), species diversity (the variety of species within ecosystems), and ecosystem diversity (the variety of habitats and ecosystems within landscapes). Each component plays a crucial role in maintaining the resilience and stability of ecosystems, allowing them to adapt to environmental changes and support life.

Next, we delve into the impacts of global warming on biodiversity, which are becoming increasingly evident in ecosystems around the world. Rising temperatures, shifting precipitation patterns, and more frequent extreme weather events are disrupting ecosystems and altering species distributions. Species that are unable to adapt or migrate to more suitable habitats may face extinction, leading to loss of biodiversity and ecosystem services.

One of the most visible impacts of global warming on biodiversity is the loss of habitats due to changes in temperature and precipitation. For example, polar ice caps are melting, leading to habitat loss for polar bears, seals, and other Arctic species. Similarly, coral reefs are experiencing mass bleaching events due to warmer ocean temperatures, leading to declines in coral cover and loss of biodiversity in marine ecosystems.

We also explore the concept of ecological thresholds, tipping points beyond which ecosystems may undergo rapid and irreversible changes. Climate change is pushing many ecosystems past their ecological thresholds, leading to regime shifts, where ecosystems transition to new states with different species compositions and functions. These shifts can have cascading effects on biodiversity and ecosystem services, with potentially profound consequences for human well-being.

But despite the challenges posed by global warming, there is still hope for biodiversity conservation. By implementing measures such as habitat restoration, sustainable land management practices, and protected area management, we can help ecosystems adapt to changing conditions and promote biodiversity conservation in a changing world. Additionally, reducing greenhouse gas emissions and mitigating climate change is essential for preserving biodiversity and ensuring the long-term health and resilience of ecosystems.

As we reflect on the impacts of global warming on biodiversity, it becomes clear that urgent action is needed to address this pressing issue. In the chapters that follow, we will explore the impacts of climate change on ecosystems, societies, and economies, and discuss strategies for mitigating and adapting to the effects of global warming.

Chapter: Ocean Acidification and its Consequences

In this chapter, we delve into one of the lesser known but equally significant consequences of global warming: ocean acidification. As atmospheric concentrations of carbon dioxide (CO_2) rise due to human activities, a significant portion of this CO_2 is absorbed by the world's oceans, leading to changes in seawater chemistry and profound impacts on marine ecosystems.

We begin by exploring the process of ocean acidification and its underlying mechanisms. When CO_2 dissolves in seawater, it reacts with water molecules to form carbonic acid, leading to a decrease in seawater pH and an increase in acidity. This process also results in the release of hydrogen ions, which can interfere with the ability of marine organisms to build and maintain their calcium carbonate shells and skeletons.

Next, we examine the impacts of ocean acidification on marine life, from microscopic plankton to large apex predators. Many marine organisms, including corals, shellfish, and certain species of algae, rely on calcium carbonate to build their shells and skeletons. However, as seawater becomes more acidic, the availability of carbonate ions decreases, making it more difficult for these organisms to form and maintain their calcium carbonate structures. This can lead to reduced growth rates, weakened shells, and increased vulnerability to predation and disease.

Ocean acidification also has far-reaching implications for marine ecosystems and the services they provide to human society. Coral reefs, often referred to as the "rainforests of the sea," are particularly vulnerable to the effects of ocean acidification. As seawater becomes more acidic, coral reefs are increasingly susceptible to bleaching events, disease outbreaks, and erosion, leading to declines in coral cover and biodiversity. This not only threatens the survival of countless marine species but also jeopardizes the livelihoods and food security of millions of people who depend on coral reefs for their livelihoods.

Furthermore, ocean acidification can have cascading effects throughout marine food webs, impacting the abundance, distribution, and behavior of marine organisms. For example, changes in the availability of carbonate ions can affect the growth and survival of plankton, which form the base of marine food chains. This, in turn, can impact the abundance of fish, seabirds, and marine mammals that rely on plankton as a food source, leading to disruptions in marine ecosystems and coastal economies.

But despite the challenges posed by ocean acidification, there is still hope for marine conservation and resilience. By reducing greenhouse gas emissions and mitigating climate change, we can slow the rate of ocean acidification and give marine organisms more time to adapt to changing conditions. Additionally, implementing measures such as marine protected areas, sustainable fisheries management, and habitat restoration can help build resilience in marine ecosystems and promote the long-term health and sustainability of our oceans.

As we reflect on the consequences of ocean acidification, it becomes clear that urgent action is needed to address this pressing issue. In the chapters that follow, we will explore the impacts of climate change on ecosystems, societies, and economies, and discuss strategies for mitigating and adapting to the effects of global warming on our oceans.

Chapter: Deforestation and Carbon Sequestration

In this chapter, we explore the critical relationship between deforestation and carbon sequestration, highlighting the profound impacts of forest loss on the global carbon cycle and climate change. Forests play a crucial role in storing carbon dioxide (CO2) from the atmosphere, helping to regulate the Earth's climate and mitigate the effects of greenhouse gas emissions. However, deforestation and forest degradation are leading to significant losses of carbon stored in forests, exacerbating global warming, and contributing to climate change.

We begin by examining the importance of forests as carbon sinks and the processes through which they sequester carbon from the atmosphere. Trees absorb CO2 during photosynthesis, using sunlight to convert carbon dioxide and water into organic matter and oxygen. This organic matter is stored in the form of biomass, including leaves, branches, roots, and soil organic matter, where it can remain for decades to centuries. By sequestering carbon in forests, trees help to remove CO2 from the atmosphere, mitigating the effects of greenhouse gas emissions and slowing the rate of global warming.

Next, we delve into the drivers of deforestation and forest degradation, which include agricultural expansion, logging, infrastructure development, and wildfires. These activities result in the loss of forest cover and the release of stored carbon into the atmosphere, contributing to climate change and biodiversity loss. Deforestation is particularly prevalent in tropical regions, where large areas of forest are cleared each year for agriculture, grazing, and urbanization.

We also explore the impacts of deforestation on the global carbon cycle and climate change. When forests are cleared or degraded, the carbon stored in trees and soil is released into the atmosphere as CO_2, contributing to the greenhouse effect and global warming. Additionally, deforestation disrupts ecosystem functions such as evapotranspiration and cloud formation, leading to changes in local and regional climate patterns, including altered rainfall patterns and increased temperatures.

But despite the challenges posed by deforestation, there is still hope for forest conservation and carbon sequestration. By implementing measures such as sustainable forest management, reforestation, and afforestation, we can protect existing forests and restore degraded landscapes, helping to conserve biodiversity and sequester carbon. Additionally, supporting initiatives such as REDD+ (Reducing Emissions from Deforestation and Forest Degradation) can provide incentives for forest conservation and sustainable land use practices, contributing to global efforts to mitigate climate change.

As we reflect on the critical role of forests in carbon sequestration, it becomes clear that urgent action is needed to address deforestation and promote sustainable forest management. In the chapters that follow, we will explore the impacts of climate change on ecosystems, societies, and economies, and discuss strategies for mitigating and adapting to the effects of global warming on our forests.

Chapter: Geology: Unraveling the Earth's Geological History

In this chapter, we embark on a journey into the fascinating world of geology, where we explore the Earth's geological history and the processes that have shaped our planet over billions of years. Geology provides us with valuable insights into the forces that have sculpted the Earth's surface, from the movement of tectonic plates to the formation of mountains, oceans, and continents.

We begin by examining the principles of geology and the methods used by geologists to unravel the Earth's geological history. By studying rock formations, fossils, and other geological features, geologists can reconstruct past environments and events, piecing together the story of the Earth's evolution over time. This process, known as stratigraphy, allows us to understand how the Earth's surface has changed over millions of years and how geological processes have shaped the landscapes we see today.

Next, we delve into the concept of plate tectonics, one of the central principles of modern geology. According to the theory of plate tectonics, the Earth's lithosphere is divided into several large and small plates that float on the semi-fluid asthenosphere beneath them. These plates are in constant motion, driven by the heat and pressure of the Earth's interior, leading to the formation of mountains, volcanoes, and earthquakes along their boundaries. Plate tectonics provides a framework for understanding the distribution of continents and oceans, as well as the geological processes that shape the Earth's surface.

We also explore the role of geology in understanding past climate change and its impacts on the Earth's surface. By studying sedimentary rocks, ice cores, and other geological records, geologists can reconstruct past climates and identify key events, such as ice ages and mass extinctions, which have shaped the course of Earth's history. This knowledge helps us understand the factors driving climate change today and provides valuable insights into how the Earth's climate system responds to external forcings.

But perhaps most importantly, geology provides us with a sense of perspective on the scale and magnitude of geological processes. While human activities such as fossil fuel extraction and deforestation can have significant local and regional impacts, they pale in comparison to the vast forces that have shaped the Earth's surface over geological time scales. By studying the Earth's geological history, we gain a deeper appreciation for the resilience and dynamism of our planet and the importance of protecting its natural systems for future generations.

As we reflect on the contributions of geology to our understanding of the Earth's history, it becomes clear that this field of science plays a crucial role in informing our response to global challenges such as climate change and environmental degradation. In the chapters that follow, we will explore the impacts of climate change on ecosystems, societies, and economies, and discuss strategies for mitigating and adapting to the effects of global warming on our planet's geological systems.

Chapter: Geological Evidence of Past Climate Change

In this chapter, we explore the wealth of geological evidence that provides insights into past climate change events, spanning millions of years of Earth's history. By studying sedimentary rocks, ice cores, fossil records, and other geological features, scientists have reconstructed past climates and identified key drivers of climate variability, shedding light on the mechanisms and consequences of natural climate change.

We begin by examining the role of sedimentary rocks as archives of Earth's climate history. Layers of sediment deposited in ancient oceans, lakes, and riverbeds contain valuable clues about past climate conditions, including temperature, precipitation, and sea level. By analyzing the composition and characteristics of sedimentary rocks, geologists can reconstruct past environments and climate regimes, from ancient tropical forests to ice-covered polar regions.

Next, we delve into the study of ice cores, which provide a unique window into past climates and atmospheric conditions. Ice cores drilled from polar ice caps and glaciers contain layers of snow and ice that accumulate over thousands to hundreds of thousands of years, preserving a record of past climates and atmospheric composition. By analyzing isotopic ratios, gas concentrations, and other proxies preserved in ice cores, scientists can reconstruct past temperatures, greenhouse gas levels, and climate variability with remarkable precision.

We also explore the fossil record as a valuable source of information about past climates and ecosystems. Fossils of plants, animals, and microorganisms preserved in sedimentary rocks provide insights into past environments, biodiversity, and evolutionary trends. By studying the distribution and characteristics of fossils, paleontologists can reconstruct past climates and ecosystems and identify periods of warming, cooling, and extinction events.

Additionally, we examine other geological features that provide evidence of past climate change, such as glacial deposits, coral reefs, and ancient shorelines. Glacial moraines, for example, mark the extent of past ice sheets and glaciers, providing evidence of past ice ages and interglacial periods. Similarly, coral reefs contain growth bands that record past sea surface temperatures and oceanic conditions, offering insights into past climate variability and ocean dynamics.

By synthesizing evidence from multiple lines of geological inquiry, scientists have constructed a comprehensive understanding of past climate change events and their impacts on Earth's ecosystems and environments. These insights help us contextualize and interpret current climate trends and provide valuable information for predicting future climate scenarios and developing strategies for climate mitigation and adaptation.

As we reflect on the geological evidence of past climate change, it becomes clear that Earth's climate is a dynamic and ever-changing system, shaped by a multitude of factors over geological time scales. In the chapters that follow, we will explore the impacts of climate change on ecosystems, societies, and economies, and discuss strategies for mitigating and adapting to the effects of global warming based on lessons learned from Earth's geological history.

Chapter: Impact of Human Activities on Geomorphology

In this chapter, we explore the significant impacts of human activities on the Earth's geomorphology, the study of landforms and the processes that shape them. While geological processes have shaped the Earth's surface over millions of years, human activities have increasingly become a dominant force driving changes to landscapes and landforms, leading to profound alterations in geomorphic processes and land surface dynamics.

We begin by examining the ways in which human activities have altered natural geomorphic processes, such as erosion, sediment transport, and soil formation. Activities such as deforestation, agriculture, mining, urbanization, and construction have led to widespread changes in land cover and land use, resulting in increased rates of erosion, sedimentation, and soil degradation. For example, deforestation removes vegetation cover, leading to increased runoff and soil erosion, while urbanization alters natural drainage patterns and increases the risk of flooding and landslides.

Next, we explore the impacts of human-induced climate change on geomorphological processes and landforms. Rising temperatures, shifting precipitation patterns, and more frequent extreme weather events associated with climate change are altering erosion rates, river discharge, and sediment transport dynamics, leading to changes in river morphology, coastal erosion, and landscape stability. For example, increased rainfall intensity can lead to higher rates of soil erosion and sedimentation, while rising sea levels can accelerate coastal erosion and land loss.

We also examine the role of human activities in shaping landforms and landscapes through engineering and construction projects. Activities such as dam construction, river channelization, and coastal engineering alter natural geomorphic processes and landforms, leading to changes in river morphology, sediment transport, and coastal erosion. While these projects may provide benefits such as flood control, water supply, and infrastructure development, they can also have unintended consequences for ecosystems, biodiversity, and geomorphic processes.

Furthermore, we explore the impacts of resource extraction activities, such as mining, quarrying, and oil and gas extraction, on geomorphology. These activities can lead to significant changes in landscapes and landforms, including land subsidence, soil erosion, and habitat destruction. Additionally, the disposal of waste materials, such as mine tailings and industrial pollutants, can further exacerbate environmental degradation and geomorphic disturbances.

By synthesizing evidence from multiple lines of inquiry, including field observations, remote sensing, and modeling studies, scientists can assess the cumulative impacts of human activities on geomorphology and develop strategies for sustainable land use and environmental management. These strategies may include measures such as land restoration, erosion control, sustainable land management practices, and the implementation of green infrastructure solutions to mitigate the impacts of human activities on geomorphology and promote landscape resilience.

As we reflect on the impact of human activities on geomorphology, it becomes clear that human actions are reshaping the Earth's surface at an unprecedented rate and scale. In the chapters that follow, we will explore the broader impacts of human activities on Earth's systems, including ecosystems, societies, and economies, and discuss strategies for mitigating and adapting to the effects of anthropogenic environmental changes.

Chapter: Geological Solutions to Climate Challenges

In this chapter, we explore the role of geological solutions in addressing the challenges posed by climate change. Geology offers a range of strategies and interventions that leverage Earth's natural processes and resources to mitigate and adapt to the impacts of global warming, providing innovative solutions to pressing environmental challenges.

We begin by examining the potential of geological carbon capture and storage (CCS) as a means of reducing greenhouse gas emissions and mitigating climate change. CCS involves capturing CO_2 emissions from industrial sources such as power plants and factories and injecting them deep underground into geological formations, where they are stored permanently in porous rock formations such as depleted oil and gas reservoirs, saline aquifers, and deep coal seams. By preventing CO_2 from entering the atmosphere, CCS helps to reduce the concentration of greenhouse gases in the atmosphere and mitigate the impacts of climate change.

Next, we explore the potential of enhanced weathering as a natural climate solution for removing CO2 from the atmosphere. Enhanced weathering involves accelerating natural weathering processes, such as the breakdown of silicate rocks, which consume CO2 during chemical reactions and release alkaline minerals that can neutralize acidic soils and ocean waters. By promoting weathering reactions through activities such as spreading crushed rocks on agricultural lands or coastal areas, we can enhance carbon sequestration and reduce the impacts of ocean acidification while also improving soil fertility and crop yields.

We also examine the role of geological resources such as geothermal energy in providing clean, renewable energy sources and reducing greenhouse gas emissions. Geothermal energy harnesses the heat stored beneath the Earth's surface to generate electricity and heat buildings, offering a reliable and sustainable alternative to fossil fuels. By tapping into Earth's natural heat reservoirs, geothermal energy can help to reduce reliance on carbon-intensive energy sources and mitigate the impacts of climate change while also providing economic benefits to local communities.

Furthermore, we explore the potential of geological engineering techniques such as coastal protection, land reclamation, and groundwater management in adapting to the impacts of climate change and sea-level rise. By harnessing Earth's natural processes and materials, we can design and implement infrastructure projects that enhance resilience to extreme weather events, mitigate coastal erosion and flooding, and safeguard critical habitats and ecosystems.

By integrating geological solutions into climate change mitigation and adaptation efforts, we can leverage Earth's natural processes and resources to address the complex challenges posed by global warming. These solutions offer innovative approaches for reducing greenhouse gas emissions, enhancing carbon sequestration, and building resilience to the impacts of climate change, helping to create a more sustainable and resilient future for generations to come.

As we reflect on the potential of geological solutions to climate challenges, it becomes clear that by harnessing Earth's natural processes and resources, we can develop effective strategies for addressing climate change and creating a more sustainable and resilient future for all. In the chapters that follow, we will explore additional strategies and interventions for mitigating and adapting to the impacts of global warming, drawing on the collective efforts of scientists, policymakers, and communities around the world.

Chapter: The Science Behind Global Warming

In this chapter, we delve into the scientific principles and processes that underpin global warming, providing a comprehensive overview of the mechanisms driving climate change and its impacts on Earth's systems.

We begin by exploring the greenhouse effect, a natural phenomenon that regulates the Earth's temperature and makes life on our planet possible. When sunlight reaches the Earth's surface, some of it is absorbed and converted into heat energy. This heat is then radiated back towards space in the form of infrared radiation. Greenhouse gases in the atmosphere, such as carbon dioxide (CO_2), methane (CH_4), and water vapor (H_2O), trap some of this outgoing radiation, preventing it from escaping into space and warming the Earth's surface. Without the greenhouse effect, the Earth would be too cold to support life as we know it.

Next, we examine the role of human activities in driving global warming through the release of greenhouse gases into the atmosphere. The burning of fossil fuels for energy, deforestation, industrial processes, and agricultural practices have led to a rapid increase in atmospheric concentrations of greenhouse gases, particularly CO_2. These gases trap heat in the atmosphere, leading to an intensification of the greenhouse effect and unprecedented warming of the planet.

We also explore the evidence of global warming from multiple lines of scientific inquiry, including temperature records, satellite observations, and climate models. Over the past century, the Earth's average surface temperature has risen significantly, with each successive decade being warmer than the last. This warming trend is accompanied by numerous other indicators of climate change, including melting ice caps and glaciers, rising sea levels, shifting precipitation patterns, and more frequent extreme weather events.

Furthermore, we examine the impacts of global warming on Earth's systems, including ecosystems, biodiversity, water resources, and human societies. Warming temperatures are driving changes in species distributions, migration patterns, and phenology, leading to shifts in ecosystems and disruptions to ecological processes. Rising sea levels are inundating coastal areas, threatening communities, and infrastructure, while changes in precipitation patterns are impacting agriculture, water supply, and food security around the world.

By synthesizing evidence from multiple scientific disciplines, we gain a comprehensive understanding of the mechanisms and impacts of global warming on Earth's systems. This knowledge serves as the foundation for developing effective strategies for mitigating and adapting to the effects of climate change, including reducing greenhouse gas emissions, transitioning to renewable energy sources, enhancing resilience to extreme weather events, and protecting vulnerable ecosystems and communities.

As we reflect on the science behind global warming, it becomes clear that urgent action is needed to address this pressing issue and safeguard the future of our planet. In the chapters that follow, we will explore the impacts of climate change on ecosystems, societies, and economies, and discuss strategies for building resilience and promoting sustainability in a changing world.

Chapter: Atmospheric Processes and Weather Patterns

In this chapter, we delve into the fascinating world of atmospheric processes and weather patterns, exploring the dynamic interactions that govern the Earth's climate and shape our day-to-day weather.

We begin by examining the composition and structure of the Earth's atmosphere, which is primarily composed of nitrogen (N_2) and oxygen (O_2), along with trace amounts of other gases such as carbon dioxide (CO_2), methane (CH_4), and water vapor (H_2O). The atmosphere is divided into several layers, including the troposphere, where weather occurs, the stratosphere, where the ozone layer is located, and the mesosphere and thermosphere, which extend into space.

Next, we explore the mechanisms driving atmospheric circulation, which play a key role in distributing heat around the globe and shaping weather patterns. Solar radiation heats the Earth's surface unevenly, leading to temperature differences between the equator and the poles and creating pressure gradients in the atmosphere. These pressure gradients drive the movement of air masses and the formation of atmospheric circulation patterns, including the Hadley, Ferrel, and Polar cells, as well as jet streams and ocean currents.

We also examine the processes that govern the formation of clouds and precipitation, which are critical components of the Earth's water cycle. Water vapor in the atmosphere condenses into clouds when it reaches its dew point temperature, forming tiny droplets or ice crystals. These droplets collide and coalesce to form larger droplets, which eventually fall to the Earth's surface as precipitation, including rain, snow, sleet, and hail. The distribution and intensity of precipitation are influenced by factors such as air temperature, humidity, atmospheric stability, and the presence of weather systems such as fronts and low-pressure systems.

Furthermore, we explore the role of atmospheric phenomena such as air masses, fronts, and storms in shaping weather patterns and producing extreme weather events. Air masses are large bodies of air with uniform temperature, humidity, and pressure characteristics, which can influence weather conditions when they collide and interact. Fronts are boundaries between air masses with contrasting properties, such as temperature and humidity, which can lead to the formation of clouds, precipitation, and changes in weather conditions. Storms, including thunderstorms, hurricanes, and tornadoes, are intense atmospheric disturbances that can produce strong winds, heavy rain, and other hazardous weather conditions.

By synthesizing knowledge from meteorology, climatology, and atmospheric science, we gain a deeper understanding of the complex interactions that govern the Earth's climate and shape our day-to-day weather. This understanding helps us predict and prepare for weather events, mitigate the impacts of extreme weather, and adapt to changing climate conditions, contributing to the resilience and sustainability of human societies and ecosystems.

As we reflect on atmospheric processes and weather patterns, it becomes clear that the Earth's climate is a dynamic and interconnected system, shaped by a multitude of factors and processes. In the chapters that follow, we will explore the impacts of climate change on weather patterns, extreme weather events, and atmospheric dynamics, and discuss strategies for building resilience and promoting sustainability in a changing climate.

Chapter: The Greenhouse Effect and its Mechanisms

In this chapter, we explore the fundamental principles of the greenhouse effect, a natural phenomenon that regulates the Earth's temperature and makes life on our planet possible. Understanding the mechanisms behind the greenhouse effect is essential for comprehending the drivers of climate change and the impacts of human activities on the Earth's climate system.

We begin by examining the basic components of the greenhouse effect. When sunlight reaches the Earth's atmosphere, some of it is absorbed by the Earth's surface, warming the surface, and converting solar energy into heat. This heat is then radiated back towards space in the form of infrared radiation. Greenhouse gases in the Earth's atmosphere, such as carbon dioxide (CO_2), methane (CH_4), water vapor (H_2O), and nitrous oxide (N_2O), absorb and re-emit some of this infrared radiation, trapping heat in the atmosphere and warming the Earth's surface. Without the greenhouse effect, the Earth's average surface temperature would be about -18°C (-0.4°F), too cold to support life as we know it.

Next, we explore the mechanisms by which greenhouse gases trap heat in the Earth's atmosphere. When infrared radiation is absorbed by greenhouse gases, it causes the molecules to vibrate and emit infrared radiation in all directions, including back towards the Earth's surface. This process, known as radiative forcing, increases the energy content of the Earth-atmosphere system, leading to an imbalance between incoming solar radiation and outgoing infrared radiation. As a result, the Earth's surface, and lower atmosphere warm up, creating the conditions necessary for life to thrive.

We also examine the role of feedback mechanisms in amplifying or dampening the effects of the greenhouse effect. Positive feedback loops, such as the ice-albedo feedback and the water vapor feedback, amplify the warming effect of greenhouse gases by reinforcing initial temperature changes. For example, as temperatures rise, ice and snow melt, reducing the Earth's surface albedo and causing more sunlight to be absorbed, further increasing temperatures. Negative feedback loops, such as the carbonate-silicate weathering feedback, act to stabilize the Earth's climate by counteracting temperature changes. For example, as temperatures rise, weathering processes accelerate, removing CO_2 from the atmosphere and cooling the Earth's surface.

Furthermore, we explore the role of human activities in enhancing the greenhouse effect and driving global warming. The burning of fossil fuels for energy, deforestation, industrial processes, and agricultural practices have led to a rapid increase in atmospheric concentrations of greenhouse gases, particularly CO_2. These human-induced emissions have intensified the greenhouse effect, leading to unprecedented warming of the Earth's surface and changes in the Earth's climate system.

By synthesizing knowledge from atmospheric science, climate modeling, and Earth system science, we gain a deeper understanding of the mechanisms behind the greenhouse effect and its implications for the Earth's climate. This understanding serves as the foundation for developing strategies to mitigate the impacts of global warming, including reducing greenhouse gas emissions, transitioning to renewable energy sources, and enhancing carbon sequestration in natural and engineered systems.

As we reflect on the greenhouse effect and its mechanisms, it becomes clear that human activities are altering the Earth's climate at an unprecedented rate and scale. In the chapters that follow, we will explore the impacts of climate change on ecosystems, societies, and economies, and discuss strategies for building resilience and promoting sustainability in a changing climate.

Chapter: Extreme Weather Events and Climate Change

In this chapter, we explore the complex relationship between extreme weather events and climate change, examining the mechanisms driving changes in weather patterns and the impacts of global warming on the frequency, intensity, and duration of extreme weather phenomena.

We begin by defining extreme weather events, which include events such as heatwaves, droughts, heavy rainfall, floods, hurricanes, tornadoes, and wildfires. These events are characterized by their rarity, severity, and significant impacts on human societies, economies, and ecosystems. While extreme weather events have occurred throughout Earth's history, there is growing evidence that human-induced climate change is exacerbating their frequency and intensity.

Next, we examine the mechanisms by which climate change influences extreme weather events. Rising temperatures, shifting precipitation patterns, changes in atmospheric circulation patterns, and sea level rise are all contributing factors to the increasing frequency and severity of extreme weather phenomena. For example, warmer temperatures increase the likelihood of heatwaves and droughts, while warmer oceans fuel the intensity of hurricanes and tropical storms. Changes in atmospheric circulation patterns can lead to persistent weather patterns, such as prolonged heatwaves or heavy rainfall events, exacerbating the impacts of extreme weather on communities and ecosystems.

We also explore the impacts of extreme weather events on human societies, economies, and ecosystems. Extreme heatwaves can lead to heat-related illnesses and deaths, particularly among vulnerable populations such as the elderly and the very young. Droughts can reduce water availability for agriculture, industry, and drinking water supplies, leading to crop failures, food shortages, and economic losses. Heavy rainfall events and floods can cause widespread damage to infrastructure, homes, and businesses, displacing communities and disrupting livelihoods. Hurricanes, tornadoes, and wildfires can destroy homes, forests, and wildlife habitats, leading to loss of life and biodiversity.

Furthermore, we examine the role of climate change attribution science in understanding the links between climate change and specific extreme weather events. While it is challenging to attribute individual events solely to climate change, scientists can use statistical analyses, climate models, and observational data to assess the influence of climate change on the likelihood and severity of extreme weather phenomena. These attribution studies provide valuable insights into the impacts of climate change on extreme weather and help policymakers, planners, and communities better prepare for and respond to future events.

By synthesizing knowledge from meteorology, climatology, and atmospheric science, we gain a deeper understanding of the complex interactions between climate change and extreme weather events. This understanding serves as the foundation for developing strategies to mitigate the impacts of extreme weather on human societies, economies, and ecosystems, including improving early warning systems, enhancing disaster preparedness and response, and implementing measures to reduce greenhouse gas emissions and limit global warming.

As we reflect on the relationship between extreme weather events and climate change, it becomes clear that urgent action is needed to address this pressing issue and build resilience to the impacts of a changing climate. In the chapters that follow, we will explore additional strategies and interventions for adapting to climate change and promoting sustainability in a changing world.

Chapter: Predictive Modeling and Climate Projections

In this chapter, we delve into the realm of predictive modeling and climate projections, exploring the methodologies and tools used to forecast future climate conditions and anticipate the impacts of global warming on Earth's systems.

We begin by examining the foundational principles of climate modeling, which involve the development of mathematical representations of the Earth's climate system based on physical laws, observed data, and computational simulations. Climate models simulate the interactions between various components of the Earth system, including the atmosphere, oceans, land surface, ice sheets, and biosphere, to project future climate conditions under different scenarios of greenhouse gas emissions and human activities.

Next, we explore the different types of climate models used by scientists to make predictions about future climate conditions. General circulation models (GCMs) are the most widely used type of climate model and simulate the dynamics of the atmosphere and oceans to predict large-scale climate patterns, such as temperature, precipitation, and atmospheric circulation. Earth system models (ESMs) are more comprehensive models that include additional components such as the carbon cycle, vegetation dynamics, and biogeochemical processes, allowing for a more holistic understanding of the Earth system and its responses to external forcings.

We also examine the process of climate scenario development, which involves the construction of plausible narratives about future socioeconomic and emissions trajectories that serve as inputs to climate models. These scenarios, known as representative concentration pathways (RCPs), are used to explore a range of possible future climate outcomes based on different assumptions about population growth, economic development, technological innovation, and policy interventions. By considering a range of scenarios, scientists can assess the uncertainty and robustness of climate projections and develop strategies for adapting to different future climate conditions.

Furthermore, we explore the limitations and uncertainties associated with climate modeling and projections. Climate models are complex and simplified representations of the Earth system, and uncertainties arise from factors such as imperfect knowledge of physical processes, natural variability, and limitations in computing power and data availability. However, despite these challenges, climate models have proven to be valuable tools for understanding past climate changes, projecting future climate scenarios, and informing climate policy and decision-making.

By synthesizing knowledge from meteorology, climatology, computer science, and Earth system science, we gain a deeper understanding of the methodologies and tools used to make climate projections and anticipate the impacts of global warming on Earth's systems. This understanding serves as the foundation for developing strategies to mitigate the impacts of climate change, adapt to changing conditions, and promote sustainability in a changing world.

As we reflect on the role of predictive modeling and climate projections in understanding and addressing climate change, it becomes clear that these tools are essential for informing policy, planning, and decision-making at local, regional, and global scales. In the chapters that follow, we will explore additional strategies and interventions for building resilience, reducing greenhouse gas emissions, and promoting sustainability in the face of a changing climate.

Chapter: Oceanic Dynamics: The Heartbeat of the Planet

In this chapter, we embark on a journey into the vast and dynamic realm of oceanic dynamics, exploring the crucial role of the oceans in regulating Earth's climate, shaping weather patterns, and sustaining life on our planet.

We begin by examining the fundamental principles of oceanic dynamics, which govern the movement and behavior of seawater across the Earth's oceans. Ocean currents, driven by a complex interplay of factors such as temperature gradients, wind patterns, and the Earth's rotation, play a central role in redistributing heat energy around the globe, shaping climate patterns, and driving weather systems. From the mighty Gulf Stream in the Atlantic Ocean to the nutrient-rich upwelling currents in the Pacific, ocean currents are the lifeblood of the planet, influencing everything from regional climates to global weather patterns.

Next, we explore the role of the oceans in regulating Earth's climate through processes such as heat storage, carbon sequestration, and the transport of moisture and heat energy. The oceans absorb and store vast amounts of heat from the sun, acting as a buffer against rapid temperature changes and helping to stabilize Earth's climate. Additionally, the oceans play a crucial role in the carbon cycle, absorbing CO_2 from the atmosphere through processes such as photosynthesis by phytoplankton and the dissolution of CO_2 in seawater. This process helps to mitigate the impacts of climate change by removing greenhouse gases from the atmosphere and storing them in the deep ocean.

We also examine the impacts of climate change on oceanic dynamics and marine ecosystems. Rising temperatures, melting ice caps, changing precipitation patterns, and ocean acidification are all affecting the behavior and composition of the oceans, leading to shifts in ocean currents, sea level rise, altered marine habitats, and changes in the distribution and abundance of marine species. These changes have significant implications for human societies, economies, and ecosystems, including impacts on fisheries, coastal communities, and marine biodiversity.

Furthermore, we explore the role of the oceans in driving weather patterns and extreme weather events. Ocean-atmosphere interactions, such as El Niño-Southern Oscillation (ENSO) and the North Atlantic Oscillation (NAO), influence weather patterns and climate variability around the globe, leading to phenomena such as droughts, floods, hurricanes, and monsoons. By understanding the complex interactions between the oceans and the atmosphere, scientists can improve weather forecasts, predict extreme weather events, and enhance our ability to adapt to changing climate conditions.

By synthesizing knowledge from oceanography, climatology, marine biology, and Earth system science, we gain a deeper appreciation for the vital role of the oceans in sustaining life on Earth and regulating the planet's climate. As the heartbeat of the planet, the oceans connect us all, shaping our climate, weather, and ecosystems, and providing essential services and resources that support human societies and biodiversity.

As we reflect on the profound influence of oceanic dynamics on the Earth's systems, it becomes clear that protecting and preserving the health of the oceans is essential for ensuring a sustainable and resilient future for generations to come. In the chapters that follow, we will explore additional strategies and interventions for conserving marine ecosystems, mitigating the impacts of climate change, and promoting sustainability in a changing world.

Chapter: Thermal Expansion and Sea Level Rise

In this chapter, we explore the phenomenon of thermal expansion and its role in driving sea level rise, one of the most significant impacts of global warming on Earth's oceans and coastal regions.

We begin by examining the basic principles of thermal expansion, which occurs when water molecules absorb heat energy and increase in volume, causing the water to expand and rise in level. As the Earth's atmosphere warms due to increased concentrations of greenhouse gases, such as carbon dioxide, heat is transferred to the oceans, causing them to absorb more heat energy and expand. This process, known as thermal expansion, contributes to the overall rise in sea levels observed around the globe.

Next, we explore the mechanisms and factors influencing thermal expansion and its contribution to sea level rise. The warming of the Earth's atmosphere due to greenhouse gas emissions leads to an increase in ocean temperatures, which in turn causes seawater to expand and occupy more space. The rate of thermal expansion varies regionally and seasonally, depending on factors such as ocean currents, wind patterns, and geographic features. Additionally, the melting of glaciers, ice caps, and polar ice sheets contributes to sea level rise by adding additional water to the oceans.

We also examine the impacts of sea level rise on coastal regions, ecosystems, and human societies. Rising sea levels lead to increased coastal erosion, inundation of low-lying areas, saltwater intrusion into freshwater sources, and loss of habitat for coastal plants and animals. Coastal communities are particularly vulnerable to the impacts of sea level rise, facing risks such as flooding, property damage, displacement, and loss of livelihoods. Additionally, infrastructure such as ports, roads, and buildings are at risk of damage from rising sea levels and storm surges, leading to economic losses and disruptions to coastal economies.

Furthermore, we explore strategies for mitigating the impacts of sea level rise and adapting to changing coastal conditions. Coastal engineering solutions such as seawalls, dikes, and beach nourishment can help protect coastal communities from flooding and erosion. Natural infrastructure such as mangrove forests, coral reefs, and wetlands also play a crucial role in buffering coastlines against the impacts of sea level rise, providing important ecosystem services such as wave attenuation, sediment stabilization, and habitat provision. Additionally, land use planning, zoning regulations, and coastal management policies can help reduce vulnerability to sea level rise and promote sustainable development in coastal areas.

By synthesizing knowledge from oceanography, climatology, coastal science, and engineering, we gain a deeper understanding of the complex interactions driving sea level rise and its impacts on coastal regions and human societies. This understanding serves as the foundation for developing strategies to mitigate the impacts of sea level rise, adapt to changing coastal conditions, and promote resilience in coastal communities.

As we reflect on the challenges and opportunities presented by sea level rise, it becomes clear that urgent action is needed to address this pressing issue and safeguard the future of coastal regions and communities. In the chapters that follow, we will explore additional strategies and interventions for adapting to sea level rise, protecting coastal ecosystems, and promoting sustainability in a changing climate.

Chapter: Ocean Currents and Climate Regulation

In this chapter, we delve into the intricate dynamics of ocean currents and their pivotal role in regulating Earth's climate. From the equator to the poles, ocean currents play a fundamental role in redistributing heat energy, shaping weather patterns, and influencing climate variability on a global scale.

We begin by exploring the driving forces behind ocean currents, which include wind patterns, temperature gradients, the Earth's rotation, and the topography of the ocean floor. Surface currents, driven primarily by wind patterns and the Coriolis effect, transport warm water from the equator towards the poles and cold water from the poles towards the equator. Deep ocean currents, on the other hand, are driven by density differences resulting from variations in temperature and salinity, transporting cold, dense water from the poles towards the equator and warm, less dense water from the equator towards the poles.

Next, we examine the role of ocean currents in redistributing heat energy around the globe and regulating Earth's climate. The Gulf Stream, for example, carries warm water from the Gulf of Mexico towards the North Atlantic, where it warms the climate of western Europe and helps to moderate temperatures in the region. The Antarctic Circumpolar Current acts as a barrier, isolating Antarctica from warmer waters to the north and playing a crucial role in regulating the Earth's climate by influencing atmospheric circulation patterns and climate variability.

We also explore the impacts of climate change on ocean currents and the potential implications for Earth's climate system. Rising temperatures, melting ice caps, and changes in precipitation patterns are affecting the density and circulation of ocean currents, leading to shifts in climate patterns and weather variability around the globe. For example, changes in the strength and position of ocean currents can influence regional climate conditions, such as the frequency and intensity of droughts, floods, and heatwaves, with implications for ecosystems, agriculture, water resources, and human societies.

Furthermore, we examine the feedback mechanisms and interactions between ocean currents and the atmosphere that influence climate variability and long-term climate trends. Ocean-atmosphere interactions, such as the El Niño-Southern Oscillation (ENSO) and the North Atlantic Oscillation (NAO), can influence weather patterns and climate variability on regional and global scales, leading to phenomena such as droughts, floods, hurricanes, and monsoons. By understanding the complex interactions between ocean currents and the atmosphere, scientists can improve climate models, enhance weather forecasts, and better predict the impacts of climate change on Earth's systems.

By synthesizing knowledge from oceanography, climatology, atmospheric science, and Earth system science, we gain a deeper appreciation for the critical role of ocean currents in regulating Earth's climate and shaping weather patterns around the globe. As the "heartbeat" of the planet, ocean currents connect distant regions, influence climate variability, and sustain life in the world's oceans. Protecting and preserving the health of ocean currents is essential for ensuring a stable and resilient climate system for future generations.

As we reflect on the intricate dynamics of ocean currents and their role in climate regulation, it becomes clear that urgent action is needed to address the impacts of climate change on ocean circulation patterns and protect the health of marine ecosystems. In the chapters that follow, we will explore additional strategies and interventions for mitigating the impacts of climate change, protecting marine biodiversity, and promoting sustainability in a changing world.

Chapter: Feedback Loops and Tipping Points

In this chapter, we explore the concept of feedback loops and tipping points in the Earth's climate system, examining how these mechanisms can amplify or dampen the effects of global warming and lead to abrupt and irreversible changes in Earth's systems.

We begin by defining feedback loops, which are self-reinforcing cycles that can either amplify or dampen changes in the climate system. Positive feedback loops amplify initial changes, leading to further warming or cooling, while negative feedback loops counteract changes, helping to stabilize the climate. One example of a positive feedback loop is the ice-albedo feedback: as temperatures rise, ice and snow melt, exposing darker surfaces that absorb more sunlight, leading to further warming and more ice melt. In contrast, the carbonate-silicate weathering feedback is an example of a negative feedback loop: as temperatures rise, chemical weathering rates increase, removing CO_2 from the atmosphere and cooling the climate.

Next, we explore the concept of tipping points, which are thresholds beyond which the climate system undergoes rapid and potentially irreversible changes. Tipping points can occur in various components of the Earth's system, including ice sheets, permafrost, ocean currents, and ecosystems. Once a tipping point is crossed, the system may undergo abrupt changes, such as the collapse of ice shelves, the release of methane from thawing permafrost, or the disruption of ocean circulation patterns, with far-reaching implications for ecosystems, societies, and economies.

We also examine the impacts of feedback loops and tipping points on the Earth's climate system and the potential implications for future climate change. As temperatures continue to rise due to human-induced greenhouse gas emissions, feedback loops may intensify, leading to more rapid and severe climate changes. Tipping points represent significant risks for abrupt and irreversible changes in Earth's systems, with potential consequences such as sea level rise, extreme weather events, ecosystem disruptions, and socioeconomic disruptions.

Furthermore, we explore strategies for mitigating the risks associated with feedback loops and tipping points and promoting resilience in the face of climate change. By reducing greenhouse gas emissions, enhancing carbon sequestration, protecting, and restoring ecosystems, and building adaptive capacity, we can reduce the likelihood of crossing tipping points and mitigate the impacts of feedback loops on Earth's systems. Additionally, early warning systems, monitoring networks, and scenario planning can help identify and prepare for potential tipping points and abrupt changes in Earth's systems.

By synthesizing knowledge from climate science, Earth system science, ecology, and resilience theory, we gain a deeper understanding of the complex interactions between feedback loops, tipping points, and climate change. This understanding serves as the foundation for developing strategies to mitigate the risks associated with feedback loops and tipping points, adapt to changing conditions, and promote resilience in the face of climate change.

As we reflect on the potential consequences of feedback loops and tipping points for Earth's systems, it becomes clear that urgent action is needed to address the root causes of climate change and build resilience in the face of uncertain and potentially abrupt changes. In the chapters that follow, we will explore additional strategies and interventions for addressing climate change, protecting ecosystems, and promoting sustainability in a changing world.

Chapter: The Human Factor: Understanding Anthropogenic Contributions

In this chapter, we examine the significant role of human activities in driving climate change and altering Earth's systems, with a focus on understanding the anthropogenic contributions to global warming and environmental degradation.

We begin by exploring the historical context of human influence on the Earth's climate and environment. While natural factors such as volcanic eruptions, solar radiation, and orbital variations have driven climate changes throughout Earth's history, the rapid and unprecedented warming observed in recent decades is primarily attributed to human activities. Since the Industrial Revolution, human societies have been burning fossil fuels for energy, clearing forests for agriculture and development, emitting greenhouse gases into the atmosphere, and altering land and water resources at an unprecedented scale.

Next, we examine the main anthropogenic contributors to climate change and environmental degradation. The burning of fossil fuels, such as coal, oil, and natural gas, for energy production is the largest source of greenhouse gas emissions, particularly carbon dioxide (CO_2), which is released when fossil fuels are burned. Deforestation and land-use changes also contribute to climate change by reducing the Earth's capacity to absorb CO_2 from the atmosphere and releasing stored carbon into the atmosphere. Additionally, industrial processes, agricultural practices, and waste management contribute to emissions of other greenhouse gases, such as methane (CH_4) and nitrous oxide (N_2O), as well as air pollutants such as particulate matter and ozone.

We also explore the impacts of human activities on Earth's systems and the environment. Climate change is leading to rising temperatures, melting ice caps, shifting precipitation patterns, rising sea levels, and more frequent and intense extreme weather events, with far-reaching consequences for ecosystems, biodiversity, water resources, and human societies. Deforestation and habitat destruction are driving loss of biodiversity, habitat fragmentation, and ecosystem degradation, with impacts on species extinction, ecosystem services, and human well-being. Pollution from industrial activities, transportation, agriculture, and waste disposal is contaminating air, water, and soil, leading to health problems, environmental degradation, and loss of ecosystem services.

Furthermore, we examine the social, economic, and ethical dimensions of anthropogenic contributions to climate change and environmental degradation. The impacts of climate change and environmental degradation are not evenly distributed, with vulnerable populations, such as low-income communities, indigenous peoples, and marginalized groups, disproportionately affected by the consequences of human activities. Additionally, the costs and benefits of addressing climate change and environmental degradation are unevenly distributed, with some individuals, industries, and countries bearing a disproportionate burden of mitigation and adaptation efforts.

By synthesizing knowledge from climate science, environmental science, social science, and ethics, we gain a deeper understanding of the human factor in driving climate change and environmental degradation. This understanding serves as the foundation for developing strategies to mitigate the impacts of human activities, transition to sustainable and equitable practices, and promote resilience in the face of climate change and environmental challenges.

As we reflect on the significant role of human activities in shaping Earth's systems, it becomes clear that urgent action is needed to address the root causes of climate change and environmental degradation and build a more sustainable and equitable future for all. In the chapters that follow, we will explore additional strategies and interventions for addressing climate change, protecting ecosystems, and promoting sustainability in a changing world.

Chapter: Fossil Fuel Combustion and Carbon Emissions

In this chapter, we delve into the critical role of fossil fuel combustion in driving carbon emissions, a primary driver of climate change and environmental degradation. We explore the processes involved in burning fossil fuels, the resulting emissions of carbon dioxide (CO_2), and the implications for Earth's climate and ecosystems.

We begin by examining the process of fossil fuel combustion, which involves the burning of coal, oil, and natural gas to generate energy for electricity, transportation, heating, and industrial processes. When fossil fuels are burned, carbon compounds stored in the fuels are oxidized, releasing CO_2 into the atmosphere as a byproduct of combustion. This process releases significant amounts of heat energy, which is used to power engines, generate electricity, and fuel industrial processes, but also results in the emission of greenhouse gases that contribute to climate change.

Next, we explore the magnitude and sources of carbon emissions from fossil fuel combustion. The combustion of fossil fuels is the largest source of CO_2 emissions globally, accounting for the majority of anthropogenic greenhouse gas emissions. The burning of coal, which is the most carbon-intensive fossil fuel, produces the highest CO_2 emissions per unit of energy generated, followed by oil and natural gas. Emissions from fossil fuel combustion occur across multiple sectors, including electricity generation, transportation, industry, residential and commercial heating, and agriculture.

We also examine the impacts of carbon emissions from fossil fuel combustion on Earth's climate and ecosystems. CO_2 is a potent greenhouse gas that traps heat in the Earth's atmosphere, leading to global warming and climate change. The accumulation of CO_2 in the atmosphere has caused temperatures to rise, ice caps and glaciers to melt, sea levels to rise, and weather patterns to become more extreme and unpredictable. Additionally, CO_2 emissions from fossil fuel combustion contribute to ocean acidification, as the oceans absorb a significant portion of the CO_2 released into the atmosphere, leading to changes in marine ecosystems and biodiversity.

Furthermore, we explore strategies for reducing carbon emissions from fossil fuel combustion and transitioning to cleaner and more sustainable energy sources. Renewable energy technologies, such as solar, wind, hydroelectric, and geothermal power, offer viable alternatives to fossil fuels, producing electricity without emitting CO_2 or other air pollutants. Energy efficiency measures, such as improved insulation, energy-efficient appliances, and transportation systems, can also reduce the demand for fossil fuels and lower carbon emissions. Additionally, policies and regulations, such as carbon pricing, emissions trading systems, renewable energy incentives, and subsidies for clean energy technologies, can help accelerate the transition away from fossil fuels and towards a low-carbon economy.

By synthesizing knowledge from energy science, climate science, environmental economics, and policy analysis, we gain a deeper understanding of the impacts of fossil fuel combustion on carbon emissions and climate change. This understanding serves as the foundation for developing strategies to mitigate carbon emissions, transition to cleaner energy sources, and promote sustainability and resilience in the face of climate change and environmental challenges.

As we reflect on the critical role of fossil fuel combustion in driving carbon emissions and climate change, it becomes clear that urgent action is needed to address this pressing issue and transition to a more sustainable and equitable energy system. In the chapters that follow, we will explore additional strategies and interventions for reducing carbon emissions, protecting ecosystems, and building a more sustainable future for generations to come.

Chapter: Land Use Changes and Urbanization

In this chapter, we delve into the profound impacts of land use changes and urbanization on Earth's ecosystems, biodiversity, and climate. We examine the processes driving land use changes, the expansion of urban areas, and the implications for environmental sustainability and human well-being.

We begin by exploring the drivers of land use changes, which include agricultural expansion, deforestation, urbanization, infrastructure development, and industrialization. As human populations grow and economies expand, there is increasing demand for land to support food production, urban development, transportation networks, and industrial activities. This has led to widespread conversion of natural ecosystems, such as forests, grasslands, wetlands, and mangroves, into croplands, pasturelands, cities, and industrial zones, with significant consequences for biodiversity, ecosystem services, and carbon storage.

Next, we examine the process of urbanization and the growth of cities as centers of population, economic activity, and infrastructure development. Urbanization is driven by factors such as population growth, rural-to-urban migration, economic development, and globalization, leading to the expansion of built-up areas, the conversion of agricultural and natural lands into urban land uses, and the fragmentation of ecosystems. Urban areas are characterized by high population densities, infrastructure networks, industrial activities, and consumption patterns, which have significant environmental impacts, including habitat loss, air and water pollution, land degradation, and greenhouse gas emissions.

We also explore the impacts of land use changes and urbanization on Earth's ecosystems, biodiversity, and climate. Conversion of natural ecosystems into agricultural lands, urban areas, and infrastructure corridors leads to habitat loss, fragmentation, and degradation, threatening the survival of species and ecosystems. Urbanization alters hydrological cycles, increases surface temperatures, and exacerbates heat island effects, leading to changes in local climates and microclimates. Land use changes also contribute to carbon emissions and climate change, as deforestation, urban expansion, and industrial activities release CO_2 and other greenhouse gases into the atmosphere, exacerbating global warming and its impacts on Earth's systems.

Furthermore, we examine strategies for promoting sustainable land use practices and urban development to mitigate the environmental impacts of land use changes and urbanization. Sustainable land management practices, such as agroforestry, sustainable agriculture, reforestation, and ecosystem restoration, can help maintain ecosystem services, conserve biodiversity, and sequester carbon. Compact and mixed-use urban planning, green infrastructure, sustainable transportation systems, and energy-efficient buildings can reduce urban sprawl, minimize environmental impacts, and enhance resilience to climate change.

By synthesizing knowledge from ecology, urban planning, geography, and environmental science, we gain a deeper understanding of the complex interactions between land use changes, urbanization, and environmental sustainability. This understanding serves as the foundation for developing strategies to promote sustainable land use practices, manage urban growth, and address the environmental challenges associated with land use changes and urbanization.

As we reflect on the profound impacts of land use changes and urbanization on Earth's ecosystems and climate, it becomes clear that urgent action is needed to transition to more sustainable and equitable land use and urban development practices. In the chapters that follow, we will explore additional strategies and interventions for promoting sustainability, conserving biodiversity, and building resilience in a changing world.

Chapter: Technological Solutions and Mitigation Strategies

In this chapter, we explore the innovative technological solutions and mitigation strategies that offer promising avenues for addressing climate change, reducing greenhouse gas emissions, and promoting sustainability in a rapidly changing world.

We begin by examining the role of renewable energy technologies in transitioning to a low-carbon economy. Solar, wind, hydroelectric, geothermal, and biomass energy offer clean and renewable alternatives to fossil fuels, generating electricity without emitting greenhouse gases or air pollutants. Advances in renewable energy technologies, such as improvements in solar panel efficiency, wind turbine design, and energy storage systems, have led to significant cost reductions and widespread deployment of renewable energy worldwide. By scaling up investments in renewable energy infrastructure and integrating renewable energy into electricity grids, we can accelerate the transition away from fossil fuels and towards a more sustainable and resilient energy system.

Next, we explore the potential of energy efficiency measures to reduce energy consumption, lower greenhouse gas emissions, and enhance resource efficiency. Energy efficiency improvements in buildings, transportation, industry, and appliances can reduce energy waste, lower operating costs, and decrease environmental impacts. Technologies such as energy-efficient lighting, smart thermostats, electric vehicles, and industrial automation offer opportunities for reducing energy demand and increasing productivity while reducing greenhouse gas emissions. By adopting energy-efficient technologies and practices, businesses, households, and governments can achieve significant energy savings and contribute to climate mitigation efforts.

We also examine the role of carbon capture and storage (CCS) technologies in reducing emissions from fossil fuel combustion and industrial processes. CCS technologies capture CO_2 emissions from power plants, factories, and other industrial sources and store them underground or utilize them for enhanced oil recovery. CCS can help reduce CO_2 emissions from large point sources, such as coal-fired power plants and cement factories and provide a transitional solution for industries with limited alternatives for decarbonization. Research and development efforts are underway to improve the efficiency, scalability, and cost-effectiveness of CCS technologies and explore new approaches, such as direct air capture, for removing CO_2 from the atmosphere.

Furthermore, we explore nature-based solutions for climate mitigation, such as afforestation, reforestation, and ecosystem restoration. Forests, wetlands, grasslands, and other natural ecosystems sequester carbon dioxide from the atmosphere through photosynthesis and store carbon in biomass, soils, and vegetation. By conserving and restoring natural ecosystems, enhancing carbon sinks, and promoting sustainable land management practices, we can enhance carbon sequestration, conserve biodiversity, and enhance ecosystem resilience to climate change. Nature-based solutions offer multiple co-benefits, including habitat restoration, water purification, flood mitigation, and climate adaptation, making them valuable tools for addressing climate change and promoting environmental sustainability.

By synthesizing knowledge from engineering, energy science, environmental science, and policy analysis, we gain a deeper understanding of the potential of technological solutions and mitigation strategies for addressing climate change and promoting sustainability. This understanding serves as the foundation for developing comprehensive and integrated approaches to climate mitigation, combining technological innovation, policy incentives, and societal engagement to achieve meaningful reductions in greenhouse gas emissions and build a more sustainable future for all.

As we reflect on the potential of technological solutions and mitigation strategies to address climate change, it becomes clear that urgent action is needed to accelerate the transition to a low-carbon economy and build resilience to the impacts of climate change. In the chapters that follow, we will explore additional strategies and interventions for adapting to climate change, conserving biodiversity, and promoting sustainability in a changing world.

Chapter: Impacts and Adaptation

In this chapter, we explore the impacts of climate change on Earth's systems, ecosystems, and human societies, as well as strategies for adapting to changing conditions and building resilience in the face of climate-related challenges.

We begin by examining the diverse and wide-ranging impacts of climate change on natural ecosystems, biodiversity, and ecosystem services. Rising temperatures, changing precipitation patterns, and extreme weather events are altering habitats, disrupting ecosystems, and threatening the survival of species worldwide. Coral reefs, mangrove forests, polar ecosystems, and other sensitive ecosystems are particularly vulnerable to the impacts of climate change, facing risks such as habitat loss, species extinction, and ecosystem collapse. These changes have profound implications for biodiversity, ecosystem functions, and the services that ecosystems provide, such as water purification, pollination, and carbon sequestration.

Next, we explore the impacts of climate change on human societies and communities, particularly those in vulnerable and marginalized regions. Rising temperatures, sea level rise, changing precipitation patterns, and extreme weather events are affecting agricultural productivity, water resources, food security, and livelihoods, particularly in developing countries and low-lying coastal areas. Climate-related disasters, such as hurricanes, floods, droughts, and heatwaves, are increasing in frequency and intensity, leading to loss of life, displacement, property damage, and economic losses. Vulnerable populations, such as women, children, indigenous peoples, and the poor, are disproportionately affected by the impacts of climate change, facing greater risks and fewer resources for adaptation.

We also examine strategies for adapting to climate change and building resilience in the face of climate-related challenges. Adaptation involves adjusting to changing climate conditions, reducing vulnerability to climate impacts, and enhancing adaptive capacity to cope with future changes. Adaptation strategies may include measures such as improving infrastructure resilience, enhancing water management systems, diversifying livelihoods, protecting natural ecosystems, and strengthening early warning systems and disaster preparedness. Community-based approaches to adaptation, such as participatory planning, local knowledge integration, and social mobilization, can help build resilience from the ground up, empowering communities to address their specific needs and priorities.

Furthermore, we explore the importance of mainstreaming adaptation into development planning, policies, and practices to ensure that adaptation efforts are integrated into broader sustainable development goals. By considering climate risks and vulnerabilities in decision-making processes, governments, businesses, and communities can identify opportunities for enhancing resilience, reducing risk, and promoting sustainable development pathways. Adaptation also requires international cooperation, financial support, and technology transfer to assist vulnerable countries and communities in adapting to climate change and building resilience to its impacts.

By synthesizing knowledge from climate science, ecology, development studies, and policy analysis, we gain a deeper understanding of the impacts of climate change on Earth's systems and human societies, as well as the importance of adaptation for building resilience and promoting sustainability. This understanding serves as the foundation for developing comprehensive and integrated approaches to adaptation, combining scientific knowledge, local expertise, and community engagement to address the complex challenges posed by climate change.

As we reflect on the impacts of climate change and the need for adaptation, it becomes clear that urgent action is needed to build resilience, protect vulnerable communities, and promote sustainability in a changing world. In the chapters that follow, we will explore additional strategies and interventions for mitigating the impacts of climate change, conserving biodiversity, and building a more sustainable future for all.

Chapter: Ecological Impacts: Disruptions to Ecosystem Services

In this chapter, we explore the ecological impacts of climate change on Earth's ecosystems, focusing on the disruptions to ecosystem services and the implications for biodiversity, human well-being, and the functioning of the planet.

We begin by examining the concept of ecosystem services, which are the benefits that ecosystems provide to humans and other organisms, including provisioning services (such as food, water, and timber), regulating services (such as climate regulation, water purification, and flood control), cultural services (such as recreation, tourism, and spiritual value), and supporting services (such as nutrient cycling, soil formation, and pollination). Ecosystem services are essential for human survival and well-being, providing the foundation for food security, clean water, climate regulation, and other essential functions that support life on Earth.

Next, we explore how climate change is disrupting ecosystem services and altering the functioning of ecosystems. Rising temperatures, changing precipitation patterns, and extreme weather events are affecting the distribution, composition, and productivity of ecosystems, leading to shifts in species distributions, changes in habitat suitability, and alterations in ecosystem processes. For example, changes in temperature and precipitation are affecting crop yields, water availability, and the timing of biological events, such as flowering, migration, and reproduction, with implications for agriculture, water resources, and biodiversity.

We also examine the impacts of climate change on specific ecosystem services and the implications for human societies and economies. Climate change is affecting the provision of food, water, and fiber from agricultural systems, fisheries, and forestry, leading to reduced productivity, increased food insecurity, and loss of livelihoods, particularly in vulnerable regions. Changes in climate patterns are also affecting the regulation of diseases, pests, and natural hazards, such as floods, droughts, and wildfires, with implications for human health, safety, and resilience. Additionally, changes in climate conditions are affecting cultural services, such as recreational opportunities, spiritual values, and aesthetic appreciation of nature, with implications for human well-being and quality of life.

Furthermore, we explore strategies for mitigating the impacts of climate change on ecosystem services and enhancing ecosystem resilience. Conservation and restoration of natural ecosystems, such as forests, wetlands, mangroves, and coral reefs, can help maintain biodiversity, enhance carbon sequestration, and provide multiple ecosystem services, such as water purification, flood control, and habitat provision. Sustainable land management practices, such as agroforestry, conservation agriculture, and integrated watershed management, can help enhance soil fertility, water retention, and crop resilience to climate change. Additionally, policies and incentives that promote ecosystem-based approaches to adaptation, such as payments for ecosystem services, biodiversity offsets, and green infrastructure investments, can help enhance ecosystem resilience and promote sustainable development in a changing climate.

By synthesizing knowledge from ecology, climate science, economics, and policy analysis, we gain a deeper understanding of the ecological impacts of climate change on ecosystem services and the importance of ecosystem-based approaches to adaptation and mitigation. This understanding serves as the foundation for developing strategies to protect and restore ecosystems, conserve biodiversity, and promote sustainable development in a changing climate.

As we reflect on the ecological impacts of climate change and the disruptions to ecosystem services, it becomes clear that urgent action is needed to address the root causes of climate change, protect vulnerable ecosystems, and promote resilience in a changing world. In the chapters that follow, we will explore additional strategies and interventions for mitigating the impacts of climate change, conserving biodiversity, and building a more sustainable future for all.

Chapter: Loss of Habitats and Species Extinction

In this chapter, we explore the profound ecological impacts of climate change on habitats and species, focusing on the loss of biodiversity and the risk of species extinction.

We begin by examining the direct and indirect ways in which climate change is driving habitat loss and degradation, leading to declines in species populations and biodiversity. Rising temperatures, changing precipitation patterns, and extreme weather events are altering the distribution, composition, and productivity of ecosystems, affecting the availability of suitable habitats for plants and animals. For example, changes in temperature and precipitation are causing shifts in the ranges of species, changes in the timing of biological events, such as flowering, migration, and reproduction, and alterations in ecosystem processes, such as nutrient cycling, pollination, and seed dispersal.

Next, we explore how habitat loss and degradation are exacerbating the risk of species extinction in a changing climate. Species with narrow habitat preferences, specialized ecological requirements, limited dispersal abilities, and restricted ranges are particularly vulnerable to the impacts of climate change, facing increased risks of extinction as their habitats become unsuitable or fragmented. Climate change is also exacerbating other threats to biodiversity, such as habitat destruction, invasive species, pollution, overexploitation, and disease, further increasing the risk of species extinction.

We also examine the implications of species extinction for ecosystems, human societies, and the functioning of the planet. Biodiversity loss can lead to declines in ecosystem services, such as food production, water purification, climate regulation, and pollination, with implications for human well-being, livelihoods, and quality of life. Additionally, species extinction can disrupt ecological processes, such as nutrient cycling, energy flow, and food webs, leading to cascading effects on ecosystem stability, resilience, and functioning. Furthermore, the loss of biodiversity reduces the genetic diversity of populations, limiting their ability to adapt to changing environmental conditions and increasing their vulnerability to extinction.

Furthermore, we explore strategies for mitigating the impacts of habitat loss and species extinction and conserving biodiversity in a changing climate. Protected areas, such as national parks, wildlife reserves, and marine protected areas, play a crucial role in conserving biodiversity, providing habitats for threatened and endangered species, and maintaining ecosystem functions and services. Habitat restoration and reforestation projects can help restore degraded ecosystems, enhance habitat connectivity, and promote species recovery. Additionally, captive breeding programs, ex-situ conservation efforts, and reintroduction initiatives can help restore populations of endangered species and prevent their extinction.

By synthesizing knowledge from ecology, conservation biology, climate science, and policy analysis, we gain a deeper understanding of the impacts of climate change on habitats and species, as well as the importance of conservation efforts for mitigating biodiversity loss and promoting ecosystem resilience. This understanding serves as the foundation for developing strategies to protect and restore habitats, conserve biodiversity, and promote sustainable development in a changing climate.

As we reflect on the loss of habitats and species extinction in a changing climate, it becomes clear that urgent action is needed to address the root causes of biodiversity loss, protect vulnerable species and ecosystems, and promote resilience in a changing world. In the chapters that follow, we will explore additional strategies and interventions for mitigating the impacts of climate change, conserving biodiversity, and building a more sustainable future for all.

Chapter: Food Security and Agricultural Challenges

In this chapter, we explore the significant challenges that climate change poses to global food security and agricultural systems, as well as strategies for adapting to changing conditions and ensuring sustainable food production for a growing population.

We begin by examining the impacts of climate change on agricultural productivity, food availability, and access to nutritious food. Rising temperatures, changing precipitation patterns, and extreme weather events are affecting crop yields, livestock productivity, and fisheries production, leading to reductions in food production and increased risks of food insecurity, particularly in vulnerable regions. Changes in temperature and precipitation are altering the suitability of agricultural lands, affecting planting, and harvesting seasons, and increasing the frequency and intensity of pests, diseases, and extreme weather events, such as droughts, floods, heatwaves, and storms.

Next, we explore the implications of climate change for global food security and the livelihoods of smallholder farmers, rural communities, and vulnerable populations. Climate-related disruptions to food production can lead to food shortages, price volatility, market instability, and nutritional deficiencies, exacerbating hunger, malnutrition, poverty, and social inequalities. Smallholder farmers, who rely on rain-fed agriculture and have limited access to resources, technologies, and support services, are particularly vulnerable to the impacts of climate change, facing increased risks of crop failures, income losses, and food insecurity.

We also examine strategies for enhancing agricultural resilience and promoting sustainable food production in a changing climate. Climate-smart agricultural practices, such as conservation agriculture, agroforestry, crop diversification, integrated pest management, and improved water management, can help build resilience to climate change, improve soil fertility, conserve water resources, and increase agricultural productivity. Additionally, investments in agricultural research and innovation, extension services, and rural infrastructure can help smallholder farmers adapt to changing conditions, access markets, and improve their livelihoods.

Furthermore, we explore the importance of building adaptive capacity and strengthening food systems to cope with climate-related challenges. Climate-resilient food systems require coordinated action across multiple sectors, including agriculture, water management, infrastructure, health, nutrition, and social protection. Policies and investments that promote sustainable land use, water conservation, disaster risk reduction, social safety nets, and access to markets and finance can help enhance food security, reduce vulnerability, and promote inclusive and equitable development.

By synthesizing knowledge from agriculture, climate science, economics, and policy analysis, we gain a deeper understanding of the challenges that climate change poses to food security and agricultural systems, as well as the opportunities for adaptation and innovation. This understanding serves as the foundation for developing comprehensive and integrated approaches to food security, combining climate-smart agricultural practices, policy interventions, and community engagement to ensure sustainable food production and livelihoods in a changing climate.

As we reflect on the challenges of ensuring food security in a changing climate, it becomes clear that urgent action is needed to address the root causes of climate change, build resilience in agricultural systems, and promote sustainable development for all. In the chapters that follow, we will explore additional strategies and interventions for mitigating the impacts of climate change, enhancing food security, and building a more sustainable future for generations to come.

Chapter: Human Health Risks and Disease Dynamics

In this chapter, we explore the complex interplay between climate change and human health, examining the various ways in which climate change affects disease dynamics, exacerbates health risks, and challenges public health systems worldwide.

We begin by examining the direct and indirect impacts of climate change on human health, including changes in temperature, precipitation patterns, extreme weather events, air quality, and vector-borne diseases. Rising temperatures and heatwaves can increase the risk of heat-related illnesses, such as heat exhaustion and heatstroke, particularly among vulnerable populations, such as the elderly, children, and outdoor workers. Changes in precipitation patterns and extreme weather events, such as floods, hurricanes, and droughts, can lead to injuries, displacement, and mental health impacts, as well as disruptions to healthcare services and infrastructure.

Next, we explore how climate change influences the transmission dynamics of infectious diseases, such as vector-borne diseases (e.g., malaria, dengue fever, Zika virus), waterborne diseases (e.g., cholera, diarrheal diseases), and foodborne diseases (e.g., salmonellosis, food poisoning). Changes in temperature, precipitation, humidity, and ecological conditions can affect the distribution, abundance, and behavior of disease vectors (e.g., mosquitoes, ticks, flies) and pathogens (e.g., viruses, bacteria, parasites), leading to shifts in disease transmission patterns, outbreaks, and epidemics. Climate change can also exacerbate other determinants of disease, such as poverty, malnutrition, population displacement, and social inequalities, further increasing vulnerability to infectious diseases.

We also examine the impacts of climate change on non-communicable diseases (NCDs), such as cardiovascular diseases, respiratory diseases, cancer, and mental health disorders. Climate-related factors, such as air pollution, allergens, extreme heat, and food and water insecurity, can exacerbate the risk of NCDs, increase morbidity and mortality rates, and strain healthcare systems and resources. Vulnerable populations, such as the elderly, children, pregnant women, and individuals with pre-existing health conditions, are particularly at risk of adverse health impacts from climate change.

Furthermore, we explore strategies for adapting to climate-related health risks and building resilience in healthcare systems and communities. Public health interventions, such as heatwave preparedness plans, early warning systems, and emergency response protocols, can help reduce the risks of heat-related illnesses and extreme weather events. Vector control measures, such as mosquito nets, insecticide spraying, and habitat management, can help prevent the spread of vector-borne diseases. Additionally, investments in healthcare infrastructure, disease surveillance, health education, and capacity-building can strengthen healthcare systems and improve the resilience of communities to climate-related health risks.

By synthesizing knowledge from epidemiology, climate science, public health, and policy analysis, we gain a deeper understanding of the complex interactions between climate change and human health, as well as the importance of adaptation measures for reducing health risks and building resilience in a changing climate. This understanding serves as the foundation for developing comprehensive and integrated approaches to climate change adaptation, combining public health interventions, environmental management, and community engagement to protect human health and well-being in a changing world.

As we reflect on the health risks posed by climate change and the challenges facing public health systems, it becomes clear that urgent action is needed to address the root causes of climate change, protect vulnerable populations, and promote health equity and resilience for all. In the chapters that follow, we will explore additional strategies and interventions for mitigating the impacts of climate change, safeguarding human health, and building a more sustainable and resilient future for generations to come.

Chapter: Socioeconomic Consequences: Addressing Inequality and Vulnerability

In this chapter, we delve into the socioeconomic consequences of climate change, focusing on the ways in which climate impacts exacerbate inequality, vulnerability, and social injustices, and exploring strategies for addressing these challenges.

We begin by examining how climate change disproportionately affects vulnerable and marginalized communities, exacerbating existing socioeconomic disparities and inequalities. Low-income communities, indigenous peoples, women, children, people with disabilities, and other marginalized groups often bear the brunt of climate impacts, facing increased risks of displacement, food insecurity, water scarcity, health impacts, and loss of livelihoods. Climate-related disasters, such as hurricanes, floods, droughts, and heatwaves, can disproportionately affect vulnerable populations, leading to loss of life, property damage, and economic losses.

Next, we explore the intersectionality of climate change with other social, economic, and political factors, such as poverty, inequality, discrimination, land tenure, governance, and access to resources. Climate impacts can exacerbate existing vulnerabilities and inequalities, deepening social injustices and undermining human rights, particularly in marginalized and disadvantaged communities. For example, land degradation, water scarcity, and loss of natural resources can exacerbate conflicts over land, water, and other resources, leading to displacement, migration, and social unrest.

We also examine strategies for addressing inequality and vulnerability in the context of climate change, promoting social justice, equity, and resilience for all. Social protection measures, such as cash transfers, food assistance, and social safety nets, can help support vulnerable populations and reduce their exposure to climate-related risks. Community-based approaches to adaptation, such as participatory planning, local knowledge integration, and social mobilization, can empower marginalized communities to address their specific needs and priorities. Additionally, policies and interventions that promote gender equality, indigenous rights, land tenure security, and inclusive governance can help build resilience and promote social justice in a changing climate.

Furthermore, we explore the importance of building partnerships and coalitions to address the root causes of inequality and vulnerability and promote collective action for climate justice. Civil society organizations, grassroots movements, indigenous peoples' groups, youth activists, and other stakeholders play a crucial role in advocating for equitable and inclusive climate policies, mobilizing resources, and holding governments and corporations accountable for their actions. By fostering collaboration, solidarity, and dialogue among diverse stakeholders, we can build momentum for transformative change and promote social justice and resilience in a changing world.

By synthesizing knowledge from sociology, development studies, political science, and policy analysis, we gain a deeper understanding of the socioeconomic consequences of climate change and the importance of addressing inequality and vulnerability in climate adaptation and mitigation efforts. This understanding serves as the foundation for developing inclusive and equitable approaches to climate action, combining social justice principles, human rights frameworks, and participatory processes to ensure that the benefits and burdens of climate change responses are distributed fairly and equitably among all members of society.

As we reflect on the socioeconomic consequences of climate change and the imperative to address inequality and vulnerability, it becomes clear that urgent action is needed to promote social justice, equity, and resilience for all. In the chapters that follow, we will explore additional strategies and interventions for building a more just, equitable, and sustainable future for generations to come.

Chapter: Displacement and Climate Refugees

In this chapter, we explore the growing phenomenon of displacement driven by climate change, examining the causes, impacts, and responses to the displacement of people due to climate-related hazards and environmental changes.

We begin by defining climate refugees as individuals or communities who are forced to leave their homes or places of habitual residence due to climate-related factors, such as sea-level rise, extreme weather events, droughts, floods, desertification, and land degradation. Climate refugees may be internally displaced within their own countries or forced to migrate across national borders in search of safety, livelihoods, and opportunities for survival. Climate-induced displacement is often driven by a combination of environmental, socioeconomic, political, and cultural factors, exacerbating existing vulnerabilities and inequalities.

Next, we explore the impacts of climate-induced displacement on affected populations, communities, and societies. Climate refugees face a myriad of challenges and risks, including loss of homes, assets, and livelihoods; disruptions to social networks and cultural traditions; exposure to violence, exploitation, and discrimination; and barriers to accessing basic services, such as food, water, shelter, healthcare, and education. Displacement can lead to social tensions, conflicts, and humanitarian crises, particularly in vulnerable and fragile contexts, such as low-lying coastal areas, small island states, and regions prone to recurrent natural disasters.

We also examine the responses to climate-induced displacement at local, national, and international levels, including policy frameworks, legal mechanisms, and institutional arrangements for protecting and assisting climate refugees. While there is no specific international legal framework governing the protection of climate refugees, existing legal instruments, such as the 1951 Refugee Convention and its 1967 Protocol, the Guiding Principles on Internal Displacement, and regional agreements, provide some protection for displaced persons, including those affected by climate change. However, gaps and challenges remain in translating legal protections into effective responses, including the lack of recognition of climate refugees in national legislation, limited access to durable solutions, and gaps in coordination and cooperation among governments, humanitarian agencies, and civil society organizations.

Furthermore, we explore strategies for addressing climate-induced displacement and building resilience in vulnerable communities. Adaptation measures, such as ecosystem-based approaches, disaster risk reduction, livelihood diversification, and social protection mechanisms, can help reduce the risks of displacement and promote the resilience of communities to climate-related hazards. Additionally, efforts to address the root causes of displacement, such as poverty, inequality, land degradation, and inadequate governance, can help prevent forced migration and support the rights and dignity of affected populations.

By synthesizing knowledge from migration studies, humanitarian assistance, climate science, and policy analysis, we gain a deeper understanding of the complex challenges posed by climate-induced displacement and the need for comprehensive and coordinated responses. This understanding serves as the foundation for developing inclusive and rights-based approaches to climate adaptation, migration governance, and humanitarian assistance, ensuring that the rights and dignity of climate refugees are upheld and protected in a changing world.

As we reflect on the plight of climate refugees and the imperative to address climate-induced displacement, it becomes clear that urgent action is needed to protect the rights and dignity of affected populations, build resilience in vulnerable communities, and promote solidarity and cooperation among nations. In the chapters that follow, we will explore additional strategies and interventions for addressing the root causes of displacement, protecting human rights, and building a more just, inclusive, and resilient future for all.

Chapter: Economic Costs of Climate Change

In this chapter, we delve into the economic impacts of climate change, examining the diverse ways in which climate-related hazards and environmental changes affect economies, industries, livelihoods, and prosperity worldwide.

We begin by exploring the direct and indirect costs of climate change on economic sectors, infrastructure, and natural resources. Rising temperatures, changing precipitation patterns, and extreme weather events can lead to physical damages to infrastructure, such as buildings, roads, bridges, and utilities, as well as loss of assets, property, and agricultural productivity. Climate-related disasters, such as hurricanes, floods, droughts, and wildfires, can cause widespread destruction, disruption to economic activities, and loss of income and wealth, particularly in vulnerable regions with limited resources and adaptive capacity.

Next, we examine the impacts of climate change on key economic sectors, including agriculture, fisheries, forestry, water resources, tourism, energy, transportation, and manufacturing. Climate-related hazards and environmental changes can affect the productivity, competitiveness, and profitability of industries, leading to reduced crop yields, loss of biodiversity, disruptions to supply chains, increased production costs, and decreased revenues. For example, changes in temperature and precipitation can affect agricultural productivity and food security, while sea-level rise and ocean acidification can impact fisheries and coastal tourism.

We also explore the macroeconomic impacts of climate change on national economies, fiscal budgets, trade balances, and public finances. Climate-related disasters and environmental changes can impose significant economic costs on governments, businesses, households, and taxpayers, including expenditures for disaster response and recovery, insurance payouts, subsidies for affected industries, and investments in adaptation measures and infrastructure upgrades. Additionally, climate change can affect investment decisions, capital allocation, credit ratings, and financial markets, leading to market volatility, investor uncertainty, and reduced economic growth prospects.

Furthermore, we examine the distributional impacts of climate change on income inequality, poverty, and social welfare. Climate-related hazards and environmental changes often disproportionately affect vulnerable and marginalized populations, such as low-income communities, indigenous peoples, women, children, and people with disabilities, exacerbating existing disparities and inequalities. The economic costs of climate change can further entrench poverty, undermine livelihoods, and erode social cohesion, leading to social tensions, conflicts, and humanitarian crises.

By synthesizing knowledge from economics, finance, development studies, and policy analysis, we gain a deeper understanding of the economic costs of climate change and the importance of proactive adaptation and mitigation measures for reducing risks and promoting resilience. This understanding serves as the foundation for developing comprehensive and integrated approaches to climate action, combining economic incentives, regulatory measures, and market mechanisms to internalize the costs of carbon emissions, promote sustainable development, and build a more resilient and inclusive economy.

As we reflect on the economic costs of climate change and the imperative to take action, it becomes clear that urgent and ambitious measures are needed to address the root causes of climate change, transition to a low-carbon economy, and build resilience in vulnerable communities. In the chapters that follow, we will explore additional strategies and interventions for mitigating the impacts of climate change, promoting sustainable development, and building a more prosperous and resilient future for all.

Chapter: Social Justice and Climate Equity

In this chapter, we delve into the intersection of social justice and climate change, examining the principles of equity, fairness, and inclusivity in climate action and exploring strategies for promoting justice and equity in climate policies and interventions.

We begin by defining social justice and climate equity as the fair and equitable distribution of the benefits and burdens of climate action, ensuring that all individuals and communities have equal opportunities to participate in decision-making processes, access resources, and benefit from climate solutions. Climate change disproportionately affects vulnerable and marginalized populations, exacerbating existing disparities and inequalities based on factors such as income, race, ethnicity, gender, age, disability, and geography. Therefore, addressing climate change requires a commitment to social justice, human rights, and equity, recognizing the interconnectedness of environmental, social, and economic systems.

Next, we explore the principles of climate justice, which emphasize the need to address historical and structural injustices, respect human rights, promote participatory governance, and prioritize the needs of the most vulnerable and marginalized communities in climate policies and interventions. Climate justice calls for transformative approaches to climate action that challenge power imbalances, promote environmental justice, and empower communities to address the root causes of climate change and build resilience to its impacts. By centering the voices and experiences of frontline communities and marginalized groups, climate justice seeks to ensure that climate policies are equitable, inclusive, and responsive to the needs and priorities of all stakeholders.

We also examine strategies for advancing social justice and climate equity in climate policies, programs, and projects. Community-based approaches to adaptation, such as participatory planning, local knowledge integration, and social mobilization, can empower marginalized communities to address their specific needs and priorities and build resilience to climate impacts. Equity-focused climate policies, such as carbon pricing with revenue redistribution, green investments in frontline communities, and just transition measures for workers in carbon-intensive industries, can help reduce inequality, promote social welfare, and ensure that the benefits of climate action are shared equitably among all members of society.

Furthermore, we explore the importance of addressing environmental racism and environmental injustices, which disproportionately affect communities of color, indigenous peoples, and low-income populations, and perpetuate systemic inequalities in exposure to environmental hazards, pollution, and climate risks. Environmental justice movements advocate for the rights of affected communities to clean air, safe water, healthy food, and a livable environment, challenging the unequal distribution of environmental burdens and advocating for policies and practices that promote equity, justice, and sustainability.

By synthesizing knowledge from sociology, environmental studies, political science, and policy analysis, we gain a deeper understanding of the intersection of social justice and climate change and the importance of advancing equity and fairness in climate action. This understanding serves as the foundation for developing inclusive and transformative approaches to climate justice, combining community empowerment, policy advocacy, and social mobilization to address the root causes of climate change and promote resilience in vulnerable communities.

As we reflect on the principles of social justice and climate equity, it becomes clear that urgent and concerted efforts are needed to build a more just, equitable, and sustainable future for all. In the chapters that follow, we will explore additional strategies and interventions for advancing climate justice, promoting equity, and building resilience in a changing world.

Chapter: Adaptation Strategies: Building Resilience in a Changing World

In this chapter, we explore the importance of adaptation strategies for building resilience to the impacts of climate change and ensuring the well-being of communities, ecosystems, and economies in a changing world.

We begin by defining adaptation as the process of adjusting to changing environmental conditions, reducing vulnerability to climate impacts, and maximizing opportunities for sustainable development. Adaptation strategies encompass a wide range of actions, measures, and interventions aimed at enhancing the resilience of individuals, communities, and ecosystems to climate-related hazards and environmental changes. Adaptation is essential for addressing the unavoidable impacts of climate change, reducing risks, and protecting lives, livelihoods, and assets.

Next, we examine the principles of effective adaptation, which include anticipatory planning, risk assessment, flexibility, inclusivity, and sustainability. Anticipatory adaptation involves identifying current and future climate risks, vulnerabilities, and opportunities, and developing proactive measures to reduce risks and enhance resilience. Risk assessment helps prioritize adaptation actions based on the likelihood and severity of climate impacts, as well as the vulnerability and capacity of affected populations. Flexibility and adaptive management enable continuous learning, innovation, and adjustment in response to changing conditions and evolving knowledge. Inclusivity ensures that adaptation strategies are participatory, equitable, and responsive to the needs and priorities of all stakeholders. Sustainability emphasizes the importance of integrating adaptation into broader development goals, policies, and practices, promoting co-benefits, synergies, and long-term resilience.

We also explore the range of adaptation options and interventions available to individuals, communities, governments, and organizations. Adaptation strategies can be categorized into various sectors, including water management, agriculture, forestry, coastal protection, infrastructure, health, disaster risk reduction, and urban planning. Examples of adaptation measures include improving water efficiency and conservation, diversifying crops, and livelihoods, restoring ecosystems and natural habitats, upgrading infrastructure to withstand extreme weather events, enhancing early warning systems and emergency response mechanisms, promoting public health interventions to address climate-related diseases, and integrating climate considerations into land use planning and development decisions.

Furthermore, we examine the importance of mainstreaming adaptation into policy-making processes, governance structures, and development plans at local, national, and international levels. Effective adaptation requires strong leadership, political commitment, institutional capacity, and financial resources to support implementation, monitoring, and evaluation of adaptation actions. Integration of adaptation into national and sub-national planning processes, sectoral policies, and development strategies can help mainstream climate considerations, promote coordination and coherence, and leverage synergies and co-benefits across different sectors and scales. Additionally, collaboration and cooperation among governments, civil society organizations, private sector actors, academia, and communities are essential for sharing knowledge, best practices, and resources, and building collective resilience to climate change.

By synthesizing knowledge from climate science, economics, governance, and policy analysis, we gain a deeper understanding of the importance of adaptation strategies for building resilience in a changing world. This understanding serves as the foundation for developing comprehensive and integrated approaches to adaptation, combining technical, institutional, and socio-economic measures to address the multidimensional challenges posed by climate change and promote sustainable development for all.

As we reflect on the imperative of adaptation in the face of climate change, it becomes clear that urgent and concerted efforts are needed to mainstream adaptation, mobilize resources, and build resilience in vulnerable communities and ecosystems. In the chapters that follow, we will explore additional strategies and interventions for advancing adaptation, reducing risks, and building a more resilient and sustainable future for generations to come.

Chapter: Sustainable Development and Green Infrastructure

In this chapter, we explore the concept of sustainable development and the role of green infrastructure in promoting environmental sustainability, economic prosperity, and social well-being in a changing world.

We begin by defining sustainable development as development that meets the needs of the present without compromising the ability of future generations to meet their own needs. Sustainable development integrates economic growth, social equity, and environmental protection, recognizing the interdependence of economic, social, and environmental systems and the importance of balancing competing priorities and objectives. Sustainable development aims to foster prosperity, reduce poverty, improve quality of life, and promote resilience to environmental and social risks.

Next, we examine the principles of sustainable development, which include intergenerational equity, intra-generational equity, precautionary principle, polluter pays principle, subsidiarity, and integration. Intergenerational equity emphasizes the responsibility of current generations to conserve and sustainably manage natural resources for the benefit of future generations. Intra-generational equity seeks to reduce disparities and inequalities within and among societies, ensuring that all individuals and communities have equal opportunities to access resources, participate in decision-making, and enjoy a high quality of life. The precautionary principle calls for taking preventive action to avoid or mitigate potential environmental and social harms, even in the absence of scientific certainty. The polluter pays principle holds those responsible for environmental degradation accountable for the costs of pollution and degradation. Subsidiarity emphasizes the importance of decentralization, local decision-making, and community empowerment in sustainable development. Integration promotes the mainstreaming of environmental considerations into economic and social policies and practices, fostering coherence, synergies, and trade-offs among different goals and objectives.

We also explore the concept of green infrastructure and its role in promoting sustainability and resilience in urban and rural landscapes. Green infrastructure refers to natural and semi-natural features, such as forests, wetlands, rivers, parks, green spaces, and green roofs, as well as engineered systems, such as green buildings, permeable pavements, and green infrastructure. Green infrastructure provides a wide range of ecosystem services, including carbon sequestration, air and water purification, flood regulation, temperature regulation, habitat provision, and recreational opportunities. Green infrastructure can help mitigate the impacts of climate change, reduce greenhouse gas emissions, improve air, and water quality, enhance biodiversity, conserve natural resources, and enhance the resilience of communities to climate-related hazards.

Furthermore, we examine strategies for promoting green infrastructure and integrating it into urban and regional planning processes, infrastructure investments, and development projects. Green infrastructure planning involves identifying and prioritizing green spaces, corridors, and networks, as well as integrating green features into built environments, transportation systems, and energy infrastructure. Green infrastructure financing mechanisms, such as green bonds, public-private partnerships, and ecosystem service payments, can help mobilize resources for green projects and leverage private sector investments. Additionally, community engagement, stakeholder collaboration, and capacity-building efforts can help raise awareness, build support, and foster ownership of green infrastructure initiatives among local residents, businesses, and institutions.

By synthesizing knowledge from environmental science, urban planning, economics, and policy analysis, we gain a deeper understanding of the importance of sustainable development and green infrastructure for promoting resilience, prosperity, and well-being in a changing world. This understanding serves as the foundation for developing holistic and integrated approaches to sustainable development, combining green infrastructure investments, policy reforms, and community engagement to build a more sustainable and resilient future for all.

As we reflect on the principles of sustainable development and the potential of green infrastructure to promote resilience and sustainability, it becomes clear that urgent and transformative action is needed to mainstream sustainability principles, promote green investments, and foster inclusive and equitable development. In the chapters that follow, we will explore additional strategies and interventions for advancing sustainable development, building green infrastructure, and promoting resilience in a changing world.

Chapter: Community-Based Adaptation Initiatives

In this chapter, we explore the significance of community-based adaptation initiatives in building resilience to climate change impacts, empowering local communities, and fostering sustainable development.

We begin by defining community-based adaptation (CBA) as a participatory and bottom-up approach to adaptation that engages local communities in identifying, planning, implementing, and monitoring adaptation actions that address their specific needs, priorities, and vulnerabilities. CBA recognizes the knowledge, experiences, and capacities of local communities as valuable assets in adapting to climate change and emphasizes the importance of inclusive decision-making processes, local ownership, and empowerment in adaptation efforts.

Next, we examine the principles and key characteristics of community-based adaptation initiatives, including participation, inclusivity, equity, flexibility, scalability, and sustainability. Participation involves involving community members, including vulnerable groups such as women, youth, indigenous peoples, and marginalized populations, in all stages of the adaptation process, from problem identification to project implementation and evaluation. Inclusivity ensures that all members of the community have equal opportunities to contribute to decision-making and benefit from adaptation actions, regardless of gender, age, ethnicity, or socio-economic status. Equity emphasizes the importance of addressing underlying social inequalities and power imbalances that exacerbate vulnerability to climate change impacts and limit access to resources and opportunities. Flexibility enables adaptation initiatives to be responsive to changing environmental conditions, evolving knowledge, and local contexts, allowing for iterative learning, experimentation, and adjustment. Scalability involves the ability to replicate successful adaptation practices and scale them up to other communities and regions, maximizing impact and reaching larger populations. Sustainability focuses on building long-term resilience by integrating adaptation into local development plans, policies, and institutions, promoting local ownership and stewardship of natural resources, and enhancing adaptive capacity and self-reliance.

We also explore examples of community-based adaptation initiatives from around the world, highlighting their diversity, innovation, and effectiveness in addressing climate change impacts and promoting sustainable livelihoods. Community-based adaptation initiatives may include measures such as rainwater harvesting, drought-resistant crop cultivation, agroforestry, soil and water conservation, sustainable land management, ecosystem restoration, disaster risk reduction, early warning systems, community-based natural resource management, and livelihood diversification. These initiatives not only help communities cope with current climate risks but also build their resilience to future uncertainties and promote sustainable development pathways.

Furthermore, we examine the enabling conditions and success factors for effective community-based adaptation, including supportive policies, institutional arrangements, capacity-building efforts, funding mechanisms, and partnerships. Governments, donors, non-governmental organizations, research institutions, and civil society organizations play important roles in providing technical support, financial resources, and capacity-building opportunities to strengthen local adaptive capacity, promote knowledge exchange, and facilitate multi-stakeholder collaboration. Additionally, legal frameworks, such as national adaptation plans, climate policies, and decentralized governance structures, can create an enabling environment for community-based adaptation by recognizing and supporting the rights and capacities of local communities to adapt to climate change.

By synthesizing knowledge from development studies, environmental management, participatory approaches, and policy analysis, we gain a deeper understanding of the importance of community-based adaptation initiatives in building resilience and promoting sustainable development. This understanding serves as the foundation for developing inclusive, context-specific, and locally driven approaches to adaptation that empower communities, enhance adaptive capacity, and foster resilience in the face of climate change.

As we reflect on the principles and practices of community-based adaptation, it becomes clear that empowering local communities and fostering partnerships are essential for building resilience and promoting sustainable development in a changing climate. In the chapters that follow, we will explore additional strategies and interventions for advancing community-based adaptation, mainstreaming adaptation into development planning, and building resilience in vulnerable communities.

Chapter: Policy Interventions and International Cooperation

In this chapter, we examine the role of policy interventions and international cooperation in addressing the challenges of climate change, promoting sustainable development, and building resilience at global, regional, and national levels.

We begin by recognizing that climate change is a global challenge that requires coordinated action by governments, international organizations, civil society, and the private sector. Policy interventions play a crucial role in shaping responses to climate change, providing incentives for mitigation and adaptation measures, and fostering cooperation and collaboration among nations.

Next, we explore the key components of effective climate policy, including mitigation policies, adaptation policies, finance mechanisms, technology transfer, capacity-building initiatives, and governance structures. Mitigation policies aim to reduce greenhouse gas emissions and limit global warming, through measures such as carbon pricing, renewable energy incentives, energy efficiency standards, and regulations on emissions from industry, transportation, and land use. Adaptation policies seek to enhance resilience to climate impacts, through measures such as climate risk assessments, early warning systems, infrastructure upgrades, ecosystem restoration, and community-based adaptation initiatives. Finance mechanisms, such as climate finance, carbon markets, and green bonds, provide financial resources for climate action, including mitigation and adaptation projects in developing countries. Technology transfer initiatives promote the diffusion of clean and sustainable technologies, such as renewable energy, low-carbon transportation, and climate-smart agriculture, to support mitigation and adaptation efforts. Capacity-building initiatives strengthen the technical, institutional, and human capacities of countries and communities to address climate change effectively. Governance structures, such as international agreements, national policies, and multilateral institutions, provide frameworks for coordination, cooperation, and accountability in climate action.

We also examine the role of international cooperation in addressing climate change, including the United Nations Framework Convention on Climate Change (UNFCCC), the Paris Agreement, and other international agreements and initiatives. The UNFCCC, adopted in 1992, provides a framework for international cooperation on climate change, with the objective of stabilizing greenhouse gas concentrations in the atmosphere at a level that prevents dangerous anthropogenic interference with the climate system. The Paris Agreement, adopted in 2015, sets out a global action plan to limit global warming to well below 2 degrees Celsius above pre-industrial levels, and to pursue efforts to limit the temperature increase to 1.5 degrees Celsius. The Agreement calls for nationally determined contributions (NDCs) from all countries, outlining their mitigation and adaptation commitments, as well as finance, technology transfer, and capacity-building support for developing countries. International initiatives, such as the Green Climate Fund, the Global Environment Facility, and the Clean Development Mechanism, provide financial and technical support for climate projects and programs in developing countries.

Furthermore, we explore the importance of mainstreaming climate considerations into broader development policies and strategies, such as poverty reduction, sustainable development goals (SDGs), disaster risk reduction, and biodiversity conservation. Integrating climate considerations into development planning can help maximize synergies and co-benefits, such as poverty alleviation, food security, health improvements, and ecosystem protection, while minimizing trade-offs and conflicts with other development objectives. Additionally, promoting coherence and alignment among different policy frameworks and sectors can enhance the effectiveness and efficiency of climate action and contribute to building resilience and promoting sustainability at multiple scales.

By synthesizing knowledge from political science, international relations, economics, and policy analysis, we gain a deeper understanding of the importance of policy interventions and international cooperation in addressing climate change and promoting sustainable development. This understanding serves as the foundation for developing and implementing effective and equitable policies and measures to address the root causes of climate change, build resilience, and create a more sustainable and resilient future for all.

As we reflect on the opportunities and challenges of policy interventions and international cooperation in addressing climate change, it becomes clear that urgent and ambitious action is needed to accelerate progress towards global climate goals, mobilize resources, and build political momentum for transformative change. In the chapters that follow, we will explore additional strategies and interventions for advancing climate action, promoting sustainability, and building resilience in a changing world.

Chapter: Looking Forward: Pathways to a Sustainable Future

In this final chapter, we turn our attention to the future and explore pathways to a sustainable and resilient world in the face of climate change and other environmental challenges.

We begin by acknowledging the urgency of the situation and the need for bold and decisive action to address the root causes of climate change, reduce greenhouse gas emissions, and build resilience to its impacts. The scientific evidence is clear: the window of opportunity to limit global warming to safe levels is narrowing, and decisive action is needed to avoid catastrophic consequences for people and the planet.

Next, we explore potential pathways to a sustainable future, based on principles of equity, justice, resilience, and sustainability. These pathways involve a transformation of our societies and economies, shifting away from fossil fuels and unsustainable practices towards renewable energy, green technologies, and sustainable lifestyles. Key elements of these pathways include:

Decarbonization: Rapidly phasing out fossil fuels and transitioning to renewable energy sources, such as solar, wind, and hydropower, to reduce greenhouse gas emissions and limit global warming.

Sustainable land use: Promoting sustainable agriculture, forestry, and land management practices that enhance carbon sequestration, protect biodiversity, and improve soil health, while ensuring food security and livelihoods for all.

Circular economy: Moving towards a circular economy model that minimizes waste, maximizes resource efficiency, and promotes reuse, recycling, and sustainable consumption patterns.

Nature-based solutions: Investing in nature-based solutions, such as ecosystem restoration, reforestation, and coastal protection, to enhance resilience to climate change, conserve biodiversity, and provide multiple benefits for people and ecosystems.

Green infrastructure: Investing in green infrastructure, such as public transportation, green buildings, and green spaces, to reduce greenhouse gas emissions, improve air and water quality, and enhance urban resilience.

Social justice and equity: Prioritizing social justice, equity, and inclusion in climate action, ensuring that the benefits and burdens of climate policies are distributed fairly and equitably among all members of society, particularly vulnerable and marginalized populations.

International cooperation: Strengthening international cooperation and solidarity to address global challenges, such as climate change, biodiversity loss, and pandemics, through multilateral agreements, partnerships, and collective action.

Education and awareness: Promoting education, awareness, and public engagement on climate change and sustainability issues, empowering individuals, and communities to act and advocate for change.

As we envision these pathways to a sustainable future, we recognize that the transition will not be easy and will require concerted efforts from governments, businesses, civil society, and individuals. However, the benefits of taking action are clear: a cleaner, healthier environment, more resilient communities, new economic opportunities, and a more equitable and just society for all.

By synthesizing knowledge from science, policy, and practice, we gain a deeper understanding of the challenges and opportunities ahead and the pathways to a sustainable future. This understanding serves as a roadmap for transformative change, inspiring us to take action and work together towards a brighter, more sustainable future for generations to come.

As we embark on this journey, let us remain hopeful and determined, knowing that by working together, we can overcome the challenges of climate change and create a more sustainable and resilient world for ourselves and future generations. The time for action is now. Let us seize this opportunity and build a better future for all.

Chapter: The Power of Collective Action: Mobilizing for Change

In this chapter, we delve into the transformative potential of collective action in addressing the pressing challenges of climate change and environmental degradation.

We begin by recognizing that individual actions, while important, are insufficient to address the scale and complexity of global environmental issues. Collective action, on the other hand, harnesses the power of communities, organizations, and movements to effect meaningful change at local, national, and global levels.

Next, we explore the principles and mechanisms of collective action, which include solidarity, collaboration, empowerment, and advocacy. Solidarity brings people together around shared values and common goals, fostering a sense of belonging and mutual support. Collaboration involves working together across sectors, disciplines, and boundaries to leverage diverse perspectives, expertise, and resources towards common objectives. Empowerment enables individuals and communities to take ownership of their futures, participate in decision-making processes, and advocate for their rights and interests. Advocacy amplifies voices, raises awareness, and mobilizes support for policy reforms, social justice, and environmental protection.

We also examine examples of successful collective action initiatives from around the world, ranging from grassroots movements and community-based organizations to global campaigns and international agreements. These initiatives demonstrate the power of collective action to drive positive change, mobilize resources, and influence policy decisions. Examples include:

Grassroots movements: Grassroots movements, such as Fridays for Future, Extinction Rebellion, and the Global Climate Strike, have mobilized millions of people worldwide to demand urgent action on climate change and environmental justice. These movements use creative tactics, such as protests, strikes, and civil disobedience, to raise awareness, build momentum, and pressure governments and corporations to take meaningful action.

Community-based organizations: Community-based organizations, such as indigenous peoples' groups, women's cooperatives, and youth networks, play a crucial role in building resilience, promoting sustainable livelihoods, and conserving natural resources. These organizations empower local communities to develop and implement adaptation and mitigation measures that are tailored to their needs and priorities.

Global campaigns: Global campaigns, such as the divestment movement, the Keep it in the Ground campaign, and the Plastic Free July initiative, mobilize people around the world to take collective action on pressing environmental issues. These campaigns raise awareness, change behaviors, and pressure governments and businesses to adopt more sustainable practices and policies.

International agreements: International agreements, such as the Paris Agreement, the Convention on Biological Diversity, and the Sustainable Development Goals, provide frameworks for collective action and cooperation among nations to address global environmental challenges. These agreements set targets, mobilize resources, and facilitate knowledge exchange and capacity-building to promote sustainable development and environmental protection.

Furthermore, we explore the role of technology and social media in facilitating collective action and amplifying voices for change. Digital platforms and social media networks enable individuals and groups to connect, organize, and mobilize on a global scale, empowering grassroots movements and catalyzing social and political change. Technology also facilitates data collection, analysis, and monitoring, enabling evidence-based decision-making and accountability in environmental governance.

By synthesizing knowledge from social movements, political science, communication studies, and environmental activism, we gain a deeper understanding of the power of collective action to drive positive change and create a more sustainable and just world. This understanding serves as a call to action, inspiring us to join forces, raise our voices, and work together towards a brighter future for all.

As we reflect on the power of collective action to mobilize for change, it becomes clear that each of us has a role to play in shaping the future we want to see. By coming together, we can harness our collective strength and creativity to overcome the challenges of climate change and environmental degradation, and build a more resilient, equitable, and sustainable world for generations to come.

Chapter: Grassroots Movements and Activism

In this chapter, we explore the vital role of grassroots movements and activism in driving social change, particularly in the context of addressing environmental issues such as climate change and biodiversity loss.

We begin by defining grassroots movements as bottom-up initiatives driven by ordinary people who come together to address common concerns, advocate for change, and mobilize collective action. Grassroots movements often emerge in response to perceived injustices, inequalities, or threats to the environment and human well-being. They operate at the local level but can have far-reaching impacts, shaping public discourse, influencing policy decisions, and catalyzing broader social and political movements.

Next, we examine the principles and values that underpin grassroots movements, including democracy, participation, inclusivity, empowerment, and solidarity. Grassroots movements are founded on principles of democratic decision-making, with decisions made collectively and transparently by members of the community. Participation is central to grassroots activism, with individuals and communities actively engaging in organizing, campaigning, and advocacy efforts. Inclusivity ensures that grassroots movements are open to all, regardless of background, identity, or affiliation, fostering diversity, tolerance, and mutual respect. Empowerment involves building the capacity and confidence of individuals and communities to take action, advocate for their rights, and effect positive change. Solidarity fosters a sense of common purpose and shared responsibility, binding individuals, and communities together in pursuit of common goals and values.

We also explore the strategies and tactics employed by grassroots movements to achieve their objectives, including community organizing, direct action, civil disobedience, advocacy campaigns, and social mobilization. Community organizing involves building relationships, networks, and coalitions within and among communities, empowering individuals, and groups to take collective action and shape their own futures. Direct action encompasses a range of tactics, such as protests, marches, blockades, sit-ins, and strikes, designed to raise awareness, disrupt business as usual, and pressure decision-makers to respond to demands. Civil disobedience involves nonviolent resistance to unjust laws or policies, often resulting in arrests or other forms of punishment, to highlight moral or ethical issues and challenge the status quo. Advocacy campaigns use various communication tools and strategies, such as petitions, letter-writing campaigns, social media, and public education initiatives, to raise awareness, build support, and influence public opinion and policy decisions. Social mobilization involves mobilizing people around shared values and common goals, leveraging social networks, cultural practices, and community resources to effect social change.

Furthermore, we examine examples of successful grassroots movements and activism in the environmental sphere, including the climate justice movement, the anti-fracking movement, and the indigenous rights movement. These movements have mobilized millions of people around the world to demand action on climate change, oppose destructive extractive industries, and defend the rights of indigenous peoples and local communities. Grassroots activists have played a key role in raising awareness, building alliances, and pressuring governments and corporations to adopt more sustainable and equitable policies and practices.

By synthesizing knowledge from sociology, political science, environmental studies, and activism studies, we gain a deeper understanding of the power of grassroots movements and activism to drive social change and promote environmental justice. This understanding serves as a call to action, inspiring us to support and participate in grassroots initiatives, amplify marginalized voices, and work together towards a more just, equitable, and sustainable world.

As we reflect on the power of grassroots movements and activism to effect change, it becomes clear that each of us has a role to play in shaping the future we want to see. By coming together, building solidarity, and taking collective action, we can challenge existing power structures, transform unjust systems, and create a more sustainable and equitable world for ourselves and future generations.

Chapter: Political Will and Policy Reform

In this chapter, we explore the crucial role of political will and policy reform in addressing pressing environmental challenges, particularly in the context of climate change, biodiversity loss, and ecosystem degradation.

We begin by defining political will as the determination and commitment of governments, policymakers, and other stakeholders to take action on environmental issues, allocate resources, and implement policies and programs to address them effectively. Political will is essential for overcoming inertia, resistance to change, and competing interests, and for mobilizing support for ambitious and transformative actions.

Next, we examine the factors that influence political will, including public opinion, electoral dynamics, leadership, economic interests, and international pressures. Public opinion plays a critical role in shaping political will, as public awareness, concern, and support for environmental issues can exert pressure on policymakers to prioritize them and take action. Electoral dynamics, such as the composition of government, party politics, and electoral cycles, can also influence political will, with changes in government often leading to shifts in environmental policies and priorities. Leadership is another key factor, as strong and visionary leaders can champion environmental causes, mobilize support, and drive policy reforms. Economic interests, such as those of the fossil fuel industry, agribusiness, and other extractive industries, can sometimes act as barriers to political will, as powerful vested interests may seek to undermine or block environmental regulations and reforms. International pressures, such as global agreements, treaties, and peer pressure, can also influence political will, encouraging governments to align their policies and actions with international norms and commitments.

We also explore strategies for building political will and driving policy reform, including advocacy, lobbying, public engagement, coalition-building, and institutional reforms. Advocacy involves raising awareness, mobilizing support, and influencing decision-makers through various channels, such as media campaigns, grassroots organizing, and direct lobbying. Lobbying entails engaging with policymakers, legislators, and government officials to shape policies, legislation, and regulations in line with environmental objectives. Public engagement involves involving citizens, stakeholders, and communities in decision-making processes, ensuring that policies are inclusive, responsive, and accountable to the needs and priorities of all. Coalition-building brings together diverse stakeholders, including civil society organizations, businesses, academia, and government agencies, to collaborate on shared goals and objectives. Institutional reforms seek to strengthen governance structures, improve transparency, accountability, and participation, and enhance the capacity of governments to address environmental challenges effectively.

Furthermore, we examine examples of successful political will and policy reform efforts in the environmental sphere, including the adoption of landmark international agreements, such as the Paris Agreement, the Convention on Biological Diversity, and the Kyoto Protocol. These agreements demonstrate the power of political will and international cooperation in mobilizing action on global environmental issues. National and sub-national policies, such as renewable energy targets, carbon pricing mechanisms, and protected area networks, also illustrate the potential for political will to drive transformative change at the local and regional levels.

By synthesizing knowledge from political science, environmental governance, advocacy, and policy analysis, we gain a deeper understanding of the importance of political will and policy reform in addressing environmental challenges and promoting sustainability. This understanding serves as a call to action, inspiring us to engage in advocacy, mobilize support, and hold policymakers accountable for bold and decisive action on environmental issues.

As we reflect on the power of political will and policy reform to drive change, it becomes clear that each of us has a role to play in shaping the future we want to see. By raising our voices, mobilizing support, and holding decision-makers accountable, we can create a more sustainable, equitable, and resilient world for ourselves and future generations.

Chapter: Corporate Responsibility and Sustainable Business Practices

In this chapter, we explore the role of corporate responsibility and sustainable business practices in addressing environmental challenges and promoting sustainability.

We begin by defining corporate responsibility as the ethical and accountable behavior of businesses towards society and the environment. Sustainable business practices are those that balance economic prosperity with social equity and environmental stewardship, aiming to create long-term value for all stakeholders, including shareholders, employees, customers, communities, and the planet.

Next, we examine the drivers of corporate responsibility, including consumer demand, investor pressure, regulatory requirements, and reputational considerations. Consumers are increasingly demanding products and services that are produced sustainably, ethically, and responsibly, driving businesses to adopt more sustainable practices to meet market demand. Investors are also increasingly considering environmental, social, and governance (ESG) factors in their investment decisions, rewarding companies that demonstrate strong corporate responsibility performance and penalizing those that do not. Regulatory requirements, such as environmental regulations, labor laws, and corporate governance standards, also play a critical role in shaping corporate behavior and incentivizing responsible practices. Reputational considerations, including brand image, trust, and public perception, can influence corporate decisions and behaviors, as companies seek to maintain their social license to operate and protect their brand reputation.

We also explore the principles and strategies of sustainable business practices, including corporate sustainability reporting, stakeholder engagement, supply chain management, and innovation. Corporate sustainability reporting involves measuring, disclosing, and transparently communicating environmental, social, and governance performance to stakeholders, providing accountability and driving continuous improvement. Stakeholder engagement involves actively involving stakeholders, such as employees, customers, suppliers, communities, and civil society organizations, in decision-making processes, ensuring that business practices are aligned with stakeholder expectations and needs. Supply chain management focuses on managing and mitigating environmental and social risks throughout the supply chain, including issues such as child labor, forced labor, environmental pollution, and human rights abuses. Innovation involves developing and implementing new technologies, products, and business models that minimize environmental impacts, enhance social benefits, and create value for all stakeholders.

Furthermore, we examine examples of companies that have adopted sustainable business practices and demonstrated corporate responsibility leadership in various industries. These companies prioritize environmental sustainability, social responsibility, and ethical business conduct, integrating sustainability into their core business strategies, operations, and culture. Examples include companies that have achieved carbon neutrality, zero waste, sustainable sourcing, fair labor practices, and community engagement, setting ambitious goals and targets to drive continuous improvement and innovation.

By synthesizing knowledge from business ethics, corporate governance, environmental management, and sustainability science, we gain a deeper understanding of the importance of corporate responsibility and sustainable business practices in addressing environmental challenges and promoting sustainability. This understanding serves as a call to action, inspiring businesses to adopt more responsible practices, consumers to support sustainable products and companies, and investors to integrate ESG factors into their investment decisions.

As we reflect on the role of corporate responsibility and sustainable business practices in shaping the future, it becomes clear that businesses have a critical role to play in advancing sustainability and creating positive social and environmental impact. By embracing corporate responsibility and adopting sustainable business practices, companies can contribute to building a more sustainable, equitable, and resilient world for current and future generations.

Chapter: Innovation and Technology: Shaping a Low-Carbon Future

In this chapter, we explore the pivotal role of innovation and technology in driving the transition to a low-carbon and sustainable future.

We begin by defining innovation as the process of developing new ideas, products, services, processes, or business models that create value and address societal needs. Technology refers to the application of scientific knowledge, tools, and techniques to solve practical problems and improve efficiency, productivity, and quality of life. Together, innovation and technology play a crucial role in unlocking new opportunities, overcoming challenges, and catalyzing transformative change in various sectors of the economy.

Next, we examine the drivers of innovation and technology in the context of addressing climate change and environmental sustainability. These drivers include market demand, policy incentives, research, and development (R&D) investments, collaboration and partnerships, and entrepreneurial leadership. Market demand for clean energy, sustainable products, and eco-friendly solutions creates opportunities for innovation and technological advancement, driving investment and entrepreneurship in green industries. Policy incentives, such as carbon pricing, renewable energy subsidies, tax incentives, and regulatory standards, provide market signals and incentives for businesses to invest in clean technologies and sustainable practices. R&D investments in renewable energy, energy efficiency, carbon capture and storage, sustainable agriculture, and other green technologies are essential for advancing scientific knowledge, developing new technologies, and scaling up innovative solutions. Collaboration and partnerships among governments, businesses, academia, research institutions, and civil society organizations facilitate knowledge exchange, technology transfer, and collective action on shared challenges. Entrepreneurial leadership and innovation ecosystems, such as incubators, accelerators, and venture capital funds, support the development and commercialization of new technologies, fostering a culture of innovation and entrepreneurship.

We also explore the various ways in which innovation and technology are driving the transition to a low-carbon and sustainable future across different sectors of the economy:

Energy transition: Innovations in renewable energy technologies, such as solar, wind, hydro, and geothermal power, are driving the transition away from fossil fuels towards clean and sustainable energy sources. Advancements in energy storage, grid integration, and smart grid technologies are enabling greater flexibility and reliability in renewable energy deployment, accelerating the transition to a decarbonized energy system.

Transportation: Innovations in electric vehicles (EVs), fuel cells, biofuels, and alternative propulsion systems are transforming the transportation sector, reducing emissions, improving air quality, and enhancing energy efficiency. Advances in autonomous vehicles, shared mobility, and transportation electrification are reshaping urban mobility patterns and reducing reliance on fossil fuels.

Industry and manufacturing: Innovations in green chemistry, materials science, and industrial processes are driving the transition towards more sustainable and circular production systems, minimizing waste, pollution, and resource depletion. Technologies such as 3D printing, robotics, and digital manufacturing are improving efficiency, reducing material usage, and enabling customized and decentralized production.

Agriculture and food systems: Innovations in precision agriculture, agroecology, vertical farming, and alternative proteins are promoting more sustainable and resilient food systems, reducing environmental impacts, enhancing resource efficiency, and increasing food security. Technologies such as blockchain, Internet of Things (IoT), and artificial intelligence (AI) are improving traceability, transparency, and supply chain resilience in the food industry.

Furthermore, we examine the role of innovation and technology in addressing other environmental challenges, such as biodiversity loss, ecosystem degradation, and pollution. Innovations in conservation biology, ecosystem restoration, and wildlife monitoring are helping to protect and restore biodiversity and ecosystems. Technologies such as remote sensing, GIS, and satellite imagery are providing valuable data and insights for conservation and environmental management.

By synthesizing knowledge from engineering, economics, environmental science, and innovation studies, we gain a deeper understanding of the transformative potential of innovation and technology in shaping a low-carbon and sustainable future. This understanding serves as a call to action, inspiring policymakers, businesses, researchers, and entrepreneurs to invest in innovation, embrace new technologies, and collaborate on solutions to address environmental challenges and promote sustainability.

As we reflect on the role of innovation and technology in shaping the future, it becomes clear that unlocking the full potential of innovation requires concerted efforts from all stakeholders, including governments, businesses, academia, and civil society. By harnessing the power of innovation and technology, we can accelerate the transition to a more sustainable, equitable, and resilient world for current and future generations.

Chapter: Renewable Energy Revolution

In this chapter, we explore the transformative potential of renewable energy in driving a global energy revolution towards sustainability and decarbonization.

We begin by defining renewable energy as energy derived from natural resources that are replenished on a human timescale, such as sunlight, wind, water, and biomass. Unlike fossil fuels, which are finite and contribute to greenhouse gas emissions and air pollution, renewable energy sources offer a clean, abundant, and sustainable alternative for meeting the world's energy needs.

Next, we examine the drivers of the renewable energy revolution, including technological advancements, cost reductions, policy support, public awareness, and climate imperatives. Technological advancements in solar photovoltaics (PV), wind turbines, hydropower, and other renewable energy technologies have led to significant improvements in efficiency, reliability, and scalability, making renewable energy increasingly competitive with fossil fuels. Cost reductions, driven by economies of scale, innovation, and learning curves, have made renewable energy sources cost-competitive with conventional energy sources in many regions of the world. Policy support, such as feed-in tariffs, renewable energy mandates, tax incentives, and carbon pricing mechanisms, has played a crucial role in driving investment and deployment of renewable energy technologies, creating market incentives for clean energy transition. Public awareness and concern about climate change, air pollution, and energy security have also increased demand for renewable energy solutions, driving consumer choices, corporate commitments, and government actions towards sustainability.

We also explore the various benefits of renewable energy, including environmental, economic, social, and geopolitical benefits. Environmental benefits include reduced greenhouse gas emissions, improved air and water quality, and conservation of natural resources, leading to cleaner and healthier environments for people and ecosystems. Economic benefits include job creation, economic growth, energy independence, and cost savings from reduced fossil fuel imports and externalities. Social benefits include improved energy access, energy equity, and energy resilience, particularly in rural and underserved communities. Geopolitical benefits include reduced geopolitical risks, conflicts, and dependencies associated with fossil fuel extraction, transportation, and trade.

Furthermore, we examine the rapid growth and expansion of renewable energy markets and industries worldwide. Solar PV and wind power, in particular, have experienced exponential growth in recent years, becoming increasingly mainstream and cost-competitive with conventional energy sources. Countries such as China, the United States, Germany, India, and Brazil have emerged as leaders in renewable energy deployment, investment, and innovation, driving global renewable energy transition and transformation. Investments in renewable energy capacity, research and development, infrastructure, and grid integration are expected to continue growing, as governments, businesses, and investors recognize the economic, environmental, and social benefits of renewable energy transition.

By synthesizing knowledge from energy economics, environmental science, policy analysis, and technology studies, we gain a deeper understanding of the renewable energy revolution and its potential to reshape the global energy landscape. This understanding serves as a call to action, inspiring policymakers, businesses, investors, and individuals to accelerate the transition to renewable energy, seize the opportunities for innovation and investment, and build a more sustainable, resilient, and equitable energy future for all.

As we reflect on the renewable energy revolution, it becomes clear that the transition to renewable energy is not only feasible but also essential for addressing climate change, promoting sustainable development, and securing a brighter future for generations to come. By embracing renewable energy solutions, we can harness the power of nature to meet our energy needs, protect our planet, and create a more prosperous and sustainable world for all.

Chapter: Carbon Capture and Storage Technologies

In this chapter, we delve into the innovative technologies of carbon capture and storage (CCS), exploring their potential to mitigate greenhouse gas emissions and combat climate change.

We begin by defining carbon capture and storage as a suite of technologies designed to capture carbon dioxide (CO2) emissions from industrial processes and power plants, transport them to storage sites, and securely store them underground to prevent their release into the atmosphere. CCS technologies offer a promising pathway for reducing CO2 emissions from large point sources, such as power plants, cement factories, steel mills, and refineries, which are responsible for a significant portion of global emissions.

Next, we examine the different components of CCS technology:

Carbon capture: Carbon capture involves capturing CO_2 emissions from industrial processes and power plants before they are released into the atmosphere. There are several capture technologies available, including post-combustion capture, pre-combustion capture, and oxy-fuel combustion. Post-combustion capture involves capturing CO_2 from flue gases using chemical solvents or adsorbents. Pre-combustion capture involves removing CO_2 from fuel gases before combustion, typically through gasification or reforming processes. Oxy-fuel combustion involves burning fuels in oxygen-rich environments to produce flue gases with high concentrations of CO_2, which can then be captured more easily.

Transportation: Once captured, CO_2 must be transported from capture sites to storage sites. Transportation methods include pipelines, ships, trucks, and trains. Pipelines are the most common method for transporting CO_2 over long distances, as they are efficient, cost-effective, and well-established in the oil and gas industry.

Storage: CO_2 storage involves injecting captured CO_2 into deep geological formations, such as depleted oil and gas reservoirs, saline aquifers, and deep coal seams, where it is stored securely underground. CO_2 can also be stored in the form of mineral carbonates through a process known as mineralization. Geological storage offers a long-term solution for permanently storing CO_2 and preventing its release into the atmosphere.

We also explore the potential benefits and challenges of CCS technology:

Benefits:

Mitigating climate change: CCS technology has the potential to significantly reduce CO2 emissions from large point sources, helping to mitigate climate change and limit global warming.
Supporting decarbonization: CCS can enable the continued use of fossil fuels, such as coal and natural gas, while reducing their environmental impact by capturing and storing CO2 emissions.
Enhancing energy security: CCS can help diversify energy sources and reduce reliance on imported fossil fuels, enhancing energy security and resilience.
Creating economic opportunities: CCS deployment can create jobs, stimulate economic growth, and attract investment in clean energy infrastructure and technology development.
Challenges:

Cost: CCS technology is currently more expensive than other mitigation options, such as renewable energy and energy efficiency measures, making it less economically competitive in many markets.
Scale and deployment: CCS deployment faces challenges related to scalability, infrastructure development, regulatory approval, and public acceptance, particularly in densely populated areas and sensitive ecosystems.
Environmental and social impacts: CCS projects must address potential environmental and social impacts, such as groundwater contamination, induced seismicity, land use conflicts, and displacement of communities.
Long-term liability: CCS storage sites must be monitored and maintained indefinitely to ensure the integrity and safety of stored CO2, raising questions about long-term liability and responsibility for storage sites.

Furthermore, we examine the current state of CCS deployment and research worldwide, highlighting notable projects, initiatives, and technological advancements in CCS technology. Countries such as Norway, Canada, the United States, and the United Kingdom have made significant investments in CCS infrastructure and research, demonstrating leadership in CCS deployment and innovation.

By synthesizing knowledge from engineering, geology, environmental science, and policy analysis, we gain a deeper understanding of the potential of CCS technology to mitigate greenhouse gas emissions and combat climate change. This understanding serves as a call to action, inspiring policymakers, industry stakeholders, and researchers to invest in CCS deployment, overcome barriers to adoption, and accelerate the transition to a low-carbon and sustainable future.

As we reflect on the promise and challenges of CCS technology, it becomes clear that CCS has the potential to play a critical role in addressing climate change and achieving global emissions reduction targets. By harnessing the power of CCS technology alongside other mitigation strategies, we can pave the way for a cleaner, greener, and more sustainable world for generations to come.

Chapter: Advances in Climate Engineering

In this chapter, we explore the emerging field of climate engineering, also known as geoengineering, and the innovative approaches being developed to mitigate the impacts of climate change.

We begin by defining climate engineering as the deliberate modification of the Earth's climate system to counteract the effects of climate change. Climate engineering encompasses a range of techniques aimed at either removing greenhouse gases from the atmosphere or reducing the amount of solar radiation reaching the Earth's surface.

Next, we examine the different categories of climate engineering techniques:

Carbon Dioxide Removal (CDR): CDR techniques focus on removing carbon dioxide (CO2) from the atmosphere, thereby reducing the concentration of greenhouse gases, and mitigating global warming. These techniques include afforestation and reforestation, ocean fertilization, direct air capture, enhanced weathering, and bioenergy with carbon capture and storage (BECCS).

Solar Radiation Management (SRM): SRM techniques aim to reduce the amount of solar radiation reaching the Earth's surface, thereby offsetting the warming effect of greenhouse gases. These techniques include stratospheric aerosol injection, marine cloud brightening, cirrus cloud thinning, and surface albedo modification.

We explore the potential benefits and risks associated with climate engineering:

Benefits:

Rapid mitigation: Climate engineering techniques have the potential to provide rapid and large-scale mitigation of climate change, complementing efforts to reduce greenhouse gas emissions.
Global applicability: Climate engineering can be applied globally and does not require international cooperation or consensus, making it potentially more feasible to implement than emission reduction measures.
Flexibility: Climate engineering allows for targeted interventions in specific regions or sectors, enabling adaptation to local climate impacts and vulnerabilities.
Risks:

Uncertainty: Climate engineering techniques carry significant uncertainties and potential side effects, including unintended consequences for ecosystems, weather patterns, and regional climates.

Ethical and social concerns: Climate engineering raises ethical and social concerns related to environmental justice, intergenerational equity, and the right to control the Earth's climate system.

Governance and regulation: Climate engineering lacks clear governance frameworks and regulations, raising questions about accountability, liability, and oversight.

Furthermore, we examine recent advances and research in climate engineering, highlighting promising developments and areas of innovation. Advances in carbon capture and storage technologies, such as direct air capture and enhanced weathering, offer new opportunities for removing CO_2 from the atmosphere and sequestering it permanently. Research into solar radiation management techniques, such as stratospheric aerosol injection and marine cloud brightening, is advancing our understanding of their potential efficacy, risks, and limitations.

By synthesizing knowledge from atmospheric science, engineering, ethics, and policy analysis, we gain a deeper understanding of the potential of climate engineering to address climate change and its associated challenges. This understanding serves as a call to action, inspiring policymakers, researchers, and civil society to engage in informed and inclusive discussions about the risks, benefits, and ethical implications of climate engineering.

As we reflect on the future of climate engineering, it becomes clear that further research, dialogue, and international cooperation are needed to assess its feasibility, safety, and potential impacts. By approaching climate engineering with caution, transparency, and a commitment to ethical principles, we can explore its potential as a tool for addressing climate change while minimizing risks to people and the planet.

Chapter: Cultivating Hope: Inspiring Hopeful Narratives and Positive Change

In this chapter, we explore the importance of cultivating hope as a catalyst for inspiring positive change in the face of environmental challenges, including climate change and biodiversity loss.

We begin by acknowledging the daunting nature of the environmental crises we face, which can often lead to feelings of fear, despair, and hopelessness. The scale and urgency of these challenges may seem overwhelming, and the consequences of inaction can be dire. However, we argue that in the midst of these challenges, cultivating hope is not only necessary but also essential for driving meaningful action and transformation.

Next, we delve into the concept of hope and its significance in the context of environmental activism and advocacy. Hope is more than just a fleeting emotion; it is a mindset, a belief in the possibility of positive change, and a commitment to action despite adversity. Hope empowers individuals and communities to envision a better future, to set ambitious goals, and to work tirelessly towards achieving them. It is the fuel that sustains us in the face of uncertainty and adversity, driving us to persevere in the pursuit of our ideals and aspirations.

We explore the role of storytelling and narrative framing in cultivating hope and inspiring action. Stories have the power to shape our perceptions, beliefs, and values, influencing how we interpret the world around us and how we envision the future. By telling stories of resilience, innovation, and collective action, we can inspire hope and mobilize people to join efforts to address environmental challenges. Positive narratives that highlight successful conservation efforts, renewable energy initiatives, and sustainable development projects can counteract feelings of despair and pessimism, offering glimpses of a brighter and more sustainable future.

Furthermore, we examine the importance of fostering a sense of agency and empowerment in individuals and communities. When people feel empowered to take action and believe that their actions can make a difference, they are more likely to engage in pro-environmental behaviors and advocate for change. Empowering individuals with knowledge, skills, and resources enables them to become agents of change in their own lives and communities, contributing to collective efforts to build a more sustainable and resilient world.

We also discuss the role of visionary leadership and bold action in cultivating hope and driving positive change. Leaders who articulate a compelling vision for a sustainable future and demonstrate courage, determination, and innovation in pursuing that vision can inspire hope and galvanize support for transformative action. By championing ambitious policies, investments, and initiatives, leaders can create momentum for change and pave the way for a more sustainable and equitable society.

By synthesizing insights from psychology, sociology, environmental studies, and activism, we gain a deeper understanding of the power of hope to drive positive change in the face of environmental challenges. This understanding serves as a call to action, inspiring individuals, organizations, and leaders to cultivate hope, tell hopeful stories, and take bold action to address climate change, biodiversity loss, and other pressing environmental issues.

As we reflect on the potential of hope to inspire positive change, it becomes clear that hope is not a passive sentiment but a transformative force for good. By nurturing hope and embracing a collective vision for a sustainable future, we can overcome adversity, mobilize resources, and create a world where people and the planet thrive together in harmony.

Chapter: Education and Public Awareness Campaigns

In this chapter, we explore the vital role of education and public awareness campaigns in fostering understanding, raising consciousness, and mobilizing action to address environmental challenges, including climate change, biodiversity loss, and pollution.

We begin by emphasizing the importance of education as a cornerstone of sustainable development and environmental stewardship. Education equips individuals with the knowledge, skills, and values needed to make informed decisions, adopt sustainable behaviors, and contribute to positive environmental outcomes. By integrating environmental education into formal curricula at all levels of schooling and providing opportunities for experiential learning and outdoor education, we can empower learners to become environmentally literate citizens capable of engaging in informed decision-making and taking action to protect the planet.

Next, we delve into the role of public awareness campaigns in raising consciousness and mobilizing action on environmental issues. Public awareness campaigns use various communication channels, such as mass media, social media, public events, and community outreach, to inform, engage, and inspire individuals and communities to adopt sustainable behaviors and support environmental initiatives. These campaigns aim to raise awareness about the causes and consequences of environmental challenges, highlight solutions and best practices, and promote behavior change at the individual, community, and societal levels.

We explore the key components of effective education and public awareness campaigns:

Clear messaging: Effective campaigns employ clear, concise, and compelling messaging that resonates with target audiences and conveys key information about environmental issues, solutions, and actions. Messages should be tailored to specific audiences and designed to evoke emotions, provoke reflection, and motivate action.

Engaging content: Engaging content, such as multimedia materials, interactive tools, and real-life stories, captivates audiences' attention and facilitates learning and behavior change. Content should be accessible, relatable, and culturally relevant, addressing diverse perspectives and experiences.

Targeted outreach: Targeted outreach strategies identify and reach key audiences, such as youth, educators, policymakers, businesses, and communities, through appropriate channels and platforms. Outreach efforts should leverage existing networks, partnerships, and influencers to maximize reach and impact.

Participation and involvement: Participation and involvement strategies encourage active engagement and ownership among target audiences, empowering them to take ownership of environmental issues and solutions. Campaigns may involve participatory activities, citizen science projects, volunteer opportunities, and community-based initiatives to foster a sense of ownership and agency.

Furthermore, we examine examples of successful education and public awareness campaigns from around the world. These campaigns have raised awareness, changed attitudes, and mobilized action on a wide range of environmental issues, from recycling and waste reduction to renewable energy adoption and conservation. Examples include the Earth Hour initiative, the Keep America Beautiful campaign, the Climate Reality Project, and the Plastic Pollution Coalition, among others.

By synthesizing insights from communication studies, psychology, environmental education, and marketing, we gain a deeper understanding of the potential of education and public awareness campaigns to inspire action and drive positive change. This understanding serves as a call to action, inspiring educators, communicators, policymakers, and activists to invest in effective education and awareness-raising efforts to build a more sustainable and resilient future for all.

As we reflect on the transformative power of education and public awareness campaigns, it becomes clear that by empowering individuals and communities with knowledge, skills, and motivation, we can mobilize collective action and create a world where people and the planet thrive together in harmony.

Chapter: The Role of Art and Culture in Climate Communication

In this chapter, we explore the powerful role of art and culture in climate communication, highlighting how creative expression can inspire empathy, foster understanding, and catalyze action on climate change.

We begin by acknowledging that climate change is not just a scientific or environmental issue; it is also a deeply cultural and social phenomenon that intersects with human values, beliefs, and identities. Art and culture provide a unique lens through which to explore and communicate the complex and interconnected dimensions of climate change, engaging hearts, and minds in ways that traditional forms of communication often cannot.

Next, we delve into the various ways in which art and culture can contribute to climate communication:

Emotional resonance: Art has the power to evoke emotions, stir imaginations, and create visceral connections with audiences. Through visual arts, literature, music, theater, film, and other creative mediums, artists can convey the urgency, beauty, and tragedy of climate change, eliciting empathy, compassion, and concern among viewers and listeners.

Narrative storytelling: Culture provides a rich tapestry of narratives, symbols, and myths that shape our understanding of the world and our place within it. Artists and storytellers can harness these cultural resources to craft compelling narratives about climate change, weaving together personal experiences, historical perspectives, and speculative futures to provoke reflection, challenge assumptions, and inspire action.

Alternative perspectives: Art offers alternative ways of seeing and knowing, challenging dominant narratives, and expanding our imagination of what is possible. Artists can offer fresh perspectives on climate change, reframing familiar issues in new and unexpected ways, and inviting audiences to reconsider their beliefs, values, and behaviors in light of environmental challenges.

Community engagement: Culture is inherently social, fostering connections and collaborations among individuals and communities. Art can serve as a catalyst for community engagement and dialogue around climate change, bringing people together to share stories, exchange ideas, and co-create solutions in a spirit of solidarity and collective action.

Furthermore, we examine examples of artists, collectives, and cultural institutions that are using their creative platforms to address climate change and promote sustainability. These examples include visual artists who create thought-provoking installations and exhibitions exploring the impacts of climate change on landscapes and communities, musicians who compose songs and performances inspired by environmental themes, writers who craft compelling stories and poetry about climate change and resilience, and filmmakers who produce documentaries and films that raise awareness and inspire action on climate-related issues.

By synthesizing insights from art history, cultural studies, psychology, and communication studies, we gain a deeper understanding of the transformative potential of art and culture in climate communication. This understanding serves as a call to action, inspiring artists, cultural practitioners, educators, and communicators to harness the power of creativity and imagination to engage diverse audiences in meaningful conversations about climate change and its impacts.

As we reflect on the role of art and culture in climate communication, it becomes clear that by tapping into our shared humanity, creativity, and cultural heritage, we can cultivate empathy, foster solidarity, and inspire collective action to address the defining challenge of our time.

Chapter: Conclusion - Recapitulation of Key Points

In this concluding chapter, we recapitulate the key points discussed throughout this book, emphasizing the urgent need for action and collaboration to address the pressing environmental challenges facing our planet.

We began by exploring the scientific evidence and understanding of climate change, emphasizing the consensus among climate scientists that human activities are driving unprecedented changes in the Earth's climate system. We examined the impacts of climate change on ecosystems, biodiversity, and human societies, highlighting the disproportionate effects on vulnerable communities and future generations.

Next, we delved into the various factors contributing to climate change, including greenhouse gas emissions from fossil fuel combustion, deforestation, industrial activities, and land use changes. We explored the interconnected nature of environmental challenges, recognizing the complex interactions between climate change, biodiversity loss, pollution, and other environmental stressors.

Throughout the book, we examined a wide range of mitigation and adaptation strategies aimed at addressing climate change and promoting sustainability. We explored the role of renewable energy, energy efficiency, carbon capture and storage, and climate engineering in reducing greenhouse gas emissions and mitigating the impacts of climate change. We also discussed the importance of ecosystem conservation, sustainable land management, and adaptation measures to enhance resilience and protect vulnerable communities and ecosystems.

Furthermore, we explored the importance of education, public awareness campaigns, and cultural engagement in fostering understanding, empathy, and action on environmental issues. We recognized the transformative potential of art, culture, and storytelling in communicating the urgency of climate change and inspiring positive change at individual, community, and societal levels.

In conclusion, we emphasize the need for collective action, political will, and international cooperation to address the root causes of climate change and build a more sustainable and resilient future for all. We recognize that while the challenges ahead are daunting, there is also cause for hope and optimism. By embracing innovation, collaboration, and a shared commitment to environmental stewardship, we can overcome adversity and create a world where people and the planet thrive together in harmony.

As we move forward, let us draw inspiration from the collective wisdom, creativity, and resilience of humanity and the natural world. Let us work together with determination and resolve to tackle the defining challenge of our time and ensure a sustainable and prosperous future for generations to come.

Chapter: Call to Action: Individual Responsibility and Collective Engagement

In this final chapter, we issue a call to action, emphasizing the crucial role of individual responsibility and collective engagement in addressing the environmental challenges we face, particularly climate change.

We begin by acknowledging that while governments, businesses, and institutions play essential roles in shaping policies and driving systemic change, individuals also have the power to make a difference through their everyday choices, actions, and advocacy efforts. Each of us has a responsibility to reduce our carbon footprint, conserve resources, and protect the environment for current and future generations.

We urge individuals to take the following actions as part of their commitment to environmental stewardship:

Reduce energy consumption: Take steps to reduce energy consumption in your home, workplace, and transportation. This can include improving energy efficiency, using renewable energy sources, and adopting sustainable transportation options such as walking, cycling, or using public transit.

Minimize waste: Reduce, reuse, and recycle materials to minimize waste generation and conserve resources. Avoid single-use plastics and opt for reusable alternatives whenever possible. Compost organic waste to divert it from landfills and reduce methane emissions.

Conserve water: Practice water conservation techniques, such as fixing leaks, using water-efficient appliances, and landscaping with native plants that require less water. Be mindful of water usage in daily activities and strive to reduce unnecessary consumption.

Support sustainable practices: Choose products and services from companies that prioritize sustainability and environmental responsibility. Support local farmers, businesses, and organizations that engage in sustainable practices and contribute to community resilience.

Advocate for policy change: Get involved in advocacy efforts to support policies and initiatives that promote sustainability, climate action, and environmental justice. Contact elected representatives, participate in community meetings, and join grassroots organizations working on environmental issues.

In addition to individual actions, we stress the importance of collective engagement and collaboration in driving systemic change and addressing the root causes of environmental challenges. We call on governments, businesses, civil society organizations, and communities to work together to:

Implement ambitious climate policies: Governments must enact policies and regulations to reduce greenhouse gas emissions, transition to renewable energy, and promote sustainable land use and conservation practices. This includes setting targets for emissions reductions, implementing carbon pricing mechanisms, and investing in clean energy infrastructure.

Foster innovation and research: Businesses and research institutions should invest in innovation, research, and development of technologies and solutions to address environmental challenges. This includes funding research on renewable energy, carbon capture and storage, sustainable agriculture, and climate adaptation strategies.

Promote environmental education and awareness: Educators, media outlets, and cultural institutions should prioritize environmental education and public awareness campaigns to raise consciousness about climate change and inspire action. This includes integrating environmental literacy into school curricula, hosting community events, and leveraging cultural platforms to communicate environmental messages.

Foster international cooperation: Countries must work together through international agreements and partnerships to address global environmental challenges, such as climate change, biodiversity loss, and pollution. This includes honoring commitments under the Paris Agreement, supporting vulnerable countries in adapting to climate impacts, and promoting equitable and sustainable development pathways.

Empower marginalized communities: Efforts to address environmental challenges must prioritize the needs and voices of marginalized communities, including indigenous peoples, low-income communities, and people of color who are disproportionately affected by environmental injustices. This includes ensuring equitable access to resources, decision-making processes, and opportunities for participation in environmental governance.

In conclusion, we emphasize that addressing environmental challenges requires collective action and shared responsibility at all levels of society. By taking individual actions, advocating for policy change, and collaborating with others, we can create a more sustainable and resilient future for ourselves and future generations.

As we embark on this collective journey towards sustainability, let us draw inspiration from the interconnectedness of all life on Earth and the shared values of stewardship, compassion, and respect for nature. Together, we have the power to create positive change and ensure a thriving planet for generations to come.

Chapter: The Imperative of Hope: Embracing a Vision of Sustainability

In this chapter, we delve into the imperative of hope as a guiding principle in embracing a vision of sustainability, even in the face of daunting environmental challenges.

We begin by acknowledging the gravity of the environmental crises we face, including climate change, biodiversity loss, pollution, and habitat destruction. These challenges can evoke feelings of fear, despair, and helplessness, leading to a sense of hopelessness about the future of our planet. However, we argue that in the midst of these challenges, hope is not only necessary but also essential for inspiring action and driving positive change.

Hope is more than just wishful thinking or blind optimism; it is an initiative-taking and resilient mindset grounded in the belief that a better future is possible and worth striving for. Hope enables us to envision a world where people and nature thrive in harmony, where communities are resilient and equitable, and where future generations can inherit a healthy and vibrant planet.

We explore the transformative power of hope in catalyzing individual and collective action towards sustainability:

Empowerment: Hope empowers individuals to take agency and initiative in shaping their own lives and communities. It instills confidence and resilience, enabling people to overcome obstacles and pursue their goals with determination and optimism.

Engagement: Hope inspires engagement and participation in environmental initiatives, movements, and campaigns. It motivates people to get involved, speak out, and take action on issues that matter to them, whether it's advocating for climate policies, supporting conservation efforts, or promoting sustainable lifestyles.

Innovation: Hope fuels innovation and creativity in developing solutions to environmental challenges. It encourages experimentation, risk-taking, and collaboration, fostering a culture of entrepreneurship and problem-solving that drives technological, social, and policy innovations towards sustainability.

Solidarity: Hope fosters solidarity and collaboration among diverse stakeholders, bridging divides and building coalitions for change. It brings people together across cultures, generations, and sectors to work towards common goals and shared visions of a sustainable future.

Furthermore, we explore the role of visioning and storytelling in cultivating hope and inspiring action towards sustainability. Visioning involves imagining and articulating a compelling vision of a sustainable future, one that embodies our collective aspirations for a thriving planet and flourishing societies. Storytelling involves sharing narratives and experiences that convey this vision, capturing imaginations, stirring emotions, and mobilizing support for transformative change.

We examine examples of visionary leaders, movements, and initiatives that embody the spirit of hope and sustainability. These examples include indigenous communities defending their lands and rights against extractive industries, youth activists mobilizing for climate action and climate justice, and cities and regions leading the transition to renewable energy and sustainable development.

In conclusion, we emphasize the imperative of hope as a guiding principle in embracing a vision of sustainability. In the face of environmental challenges, hope serves as a beacon of light, guiding us towards a better future and inspiring us to act with courage, compassion, and determination. By embracing hope and working together towards a shared vision of sustainability, we can create a world where people and the planet thrive together in harmony, now and for generations to come.

Appendix: Glossary of Key Terms

This glossary provides definitions of key terms used throughout the book to enhance understanding of environmental concepts and terminology.

Climate Change: Refers to long-term changes in temperature, precipitation, and other atmospheric conditions on Earth, primarily driven by human activities such as burning fossil fuels and deforestation.

Biodiversity: The variety of life forms, including plants, animals, and microorganisms, and the ecosystems in which they occur. Biodiversity encompasses genetic diversity, species diversity, and ecosystem diversity.

Greenhouse Gases (GHGs): Gases that trap heat in the Earth's atmosphere, contributing to the greenhouse effect and global warming. Examples include carbon dioxide (CO_2), methane (CH_4), nitrous oxide (N_2O), and fluorinated gases.

Renewable Energy: Energy derived from sources that are replenished naturally, such as sunlight, wind, water, and biomass. Renewable energy sources are sustainable and emit fewer greenhouse gases compared to fossil fuels.

Carbon Capture and Storage (CCS): A suite of technologies designed to capture carbon dioxide (CO_2) emissions from industrial processes and power plants, transport them to storage sites, and securely store them underground to prevent their release into the atmosphere.

Sustainability: The ability to meet the needs of the present generation without compromising the ability of future generations to meet their own needs. Sustainability involves balancing environmental, social, and economic considerations to ensure long-term well-being and resilience.

Climate Resilience: The capacity of communities, ecosystems, and societies to withstand and recover from the impacts of climate change, including extreme weather events, sea level rise, and changing precipitation patterns.

Environmental Justice: The fair treatment and meaningful involvement of all people, regardless of race, ethnicity, income, or location, in environmental decision-making and the distribution of environmental benefits and burdens.

Carbon Footprint: The total amount of greenhouse gases emitted directly or indirectly by an individual, organization, product, or activity, expressed as carbon dioxide equivalent (CO2e) emissions.

Mitigation: Actions taken to reduce or prevent greenhouse gas emissions and address the root causes of climate change. Mitigation measures include transitioning to renewable energy, improving energy efficiency, and implementing carbon capture and storage technologies.

Adaptation: Actions taken to adjust to the impacts of climate change and build resilience to its effects. Adaptation measures include strengthening infrastructure, protecting natural ecosystems, and developing early warning systems for extreme weather events.

Conservation: The sustainable use and management of natural resources to preserve biodiversity, protect ecosystems, and maintain ecological balance. Conservation efforts aim to prevent habitat destruction, species extinction, and loss of ecosystem services.

Ecological Footprint: A measure of the environmental impact of human activities, expressed as the amount of biologically productive land and water area required to sustainably support those activities.

Sustainability Transition: The process of shifting towards a more sustainable and equitable society, characterized by reduced resource consumption, increased environmental stewardship, and improved well-being for all.

Environmental Education: The process of providing knowledge, skills, and values to individuals and communities to raise awareness about environmental issues, promote sustainability, and inspire action for positive change.

This glossary serves as a reference guide to key terms and concepts discussed in the book, enhancing understanding, and facilitating engagement with environmental topics and sustainability initiatives.

www.ingramcontent.com/pod-product-compliance
Lightning Source LLC
Chambersburg PA
CBHW052158220526
45471CB00004B/1723